SCIENCE AND TECHNOLOGY
VITAL NATIONAL RESOURCES

Edited by
RALPH SANDERS

Lomond Books

Mt. Airy, Maryland

1975

Library of Congress Catalog Card Number: 74-175-62

ISBN:
    Clothbound   0-912-338-11-3
    Microfiche   0-912-338-12-1

Lomond Systems, Inc.
Mt. Airy, Maryland 21771
Printed in the United States of America

*This volume is dedicated
to Fred R. Brown, a keen mind and good friend*

Contributors

Ivan Asay, formerly with the National Bureau of Standards and the Agency for International Development, Washington, D.C.

Morris C. Leikind, Chief Consultant, Library of Medical History, Sackler School of Medicine, University of Tel Aviv, Israel

Krishnan D. Mathur, Professor of History, Federal City College, Washington, D.C.

Wyndham Miles, Historian, the National Institutes of Health, Bethesda, Md.

Ralph Sanders, Professor of Public Administration, Industrial College of the Armed Forces and Adjunct Professor, The American University, Washington, D.C.

Preface

This volume grew out of an effort to design readings in the field of science and technology to fit the needs of the Industrial College of the Armed Forces, senior military educational institution under the Joint Chiefs of Staff, dedicated to the study of the management of national security. The treatment in this volume, however, does not focus on the military aspects of science and technology, but provides a general overview of the nature, contributions, and problems associated with the roles of science and technology in our national life.

This ready reference will enable a reader to gain in short order a wide knowledge relating to the nature and application of science and technology as well as to their sociological and political environment. Consequently, by design, the writing is succinct, providing the key points of the subject without encumbering them with unessential elaboration. Therefore, this book should prove helpful, not only to students of this specific subject, but also to the general reader and to those in akin fields who wish to acquire this knowledge but cannot undertake voluminous reading.

Table of Contents

Acknowledgments

I want to thank Lt. Gen. Walter J. Woolwine, U.S.A., Commandant of the Industrial College, for his encouragement during the preparation of this volume. I owe a special debt to Dr. Fred R. Brown, formerly of the Industrial College faculty and now with the Washington Center of the University of Southern California, who played a crucial role in the early design and development of the manuscript. Mrs. Frances R. Burdette of the Industrial College staff read the entire manuscript and made profitable comments. The Department of the Army merits my profound gratitude for permitting me to seek open publication of this volume.

I am deeply indebted to the National Institutes of Health, the National Bureau of Standards, and the National Science Foundation for their assistance in preparing this text. The valuable aid of Bodo Bartocha and Thomas Owen, formerly of the National Science Foundation, of Richard Hewlett of the Atomic Energy Commission, and of Frank W. Anderson of the National Aeronautics and Space Administration is especially appreciated.

I also want to thank Dr. Lowell Hattery and the administration of the School of Government and Public Administration of The American University. The university provided me with the opportunity to teach courses in science, technology, and government over many years, thereby enabling me to formulate many of my ideas on this subject.

As always, my wife made her unique contributions, not only as copy editor of the entire manuscript, but also as an intellectual sounding board that unearthed my mistaken starts as well as my final destinations.

Above all, I am grateful to the authors who contributed chapters to this volume. They did a commendable job of presenting prolix and complex subject matter in a succinct and readable manner.

The views expressed in this book are those of the editor and the contributors and do not reflect official policy of the Industrial College of the Armed Forces, the Department of Defense, or other agencies of the U.S. government.

Ralph Sanders
Professor of Public Administration

Industrial College of the Armed Forces
Washington, D.C.
1 June 1974

Introduction

Science and technology are vital national resources; no one really could argue with this premise. Over the years Americans have been inundated with discourses on the roles of science and technology in revolutionizing society. Truly the swift transformation which science and technology have initiated within our society in recent decades are demonstrated on every hand.

Yet, for the sake of balance, it would be helpful to keep in mind the old conflict in Greek philosophy expressed most clearly by the contrast between Heraclitus' doctrine of endless change and Plato's emphasis on the abiding. Heraclitus said that all things flow; nothing abides. Plato stressed immutable, lasting truths. We can agree with Heraclitus that science and technology have transformed much of the world, but we also can agree with Plato that many things remain unchanged. For example, our concept of the uniqueness of the human soul, our faith in the inalienable rights of man, and our belief in the right of self-defense remain as valid today as they did centuries ago.

Science and technology have created unprecedented situations in society; witness, for example, the fantastic lessening in time and distance. It is singular to our time that nations can be destroyed within hours. Yet, the same cooperation and competition that normally envelop human activity also enwrap science and technology. Many people seek to promote accord in these activities; programs of international scientific cooperation are one approach. At the same time, ever-present conflicts of interest have prompted competition. Today, one such conflict centers around rival claims for federal resources by basic researchers and applied developers. This struggle, of course, is a manifestation of traditional bidding by rivals for limited resources.

In military matters, science and technology have altered the way men fight wars, but they certainly have not obviated the need for leaders who engender confidence and courage in time of stress. We should not become so mesmerzied by scientific and tchnological change that we fail to recognize and appreciate the lasting phenomena of society.

Nor should we become overawed by the problems which science and technology create. Some people believe that science and technology trigger more problems than they solve. Insecticides, this argument contends, help save crops, but also pollute our rivers. DDT helps cut down death rates, but also assists the population explosion. Are we merely compounding our vexations by trying to use science and technology to reduce them?

Certainly science and technology have changed our environment and consequently our problems. One could acknowledge that they have made life more complex. But there is a vast difference in human problems between their complexity and their gravity.

Furthermore, society has developed sciences and technologies to help it cope with complexity beyond anything ever imagined by our forefathers. The long and short of this question seems to be that science and technology solve some problems, but create others. In specific cases we can analyze the resulting relationship, but we can't find any universal truth as to the weight of modern problems compared to problems throughout history.

It is no more true to say that all else is subordinate to science and technology than it is to say that science and technology do more to intensify human problems than to remedy them. Those who argue such a case suffer from what philosophers call "reductionism"—that is, one factor is fundamental and all others are derivative. More accurately, science and technology have a multidimensional, reciprocal relationship with society, constantly acting and reacting at all levels of social organization. Science and technology do not emerge from a social vacuum, but rather are influenced by social attitudes.

Another premise of this book is that a concise, yet comprehensive treatment of major aspects of a subject—as profound and complex as science and technology—can give the student useful insights into their nature and background. The book is designed to limit discussion to highlights while providing selected bibliography for those who desire to explore any specific subject in greater depth.

Chapter 1, "The Nature and Meaning of Science and Technology," discusses the fundamentals and social role of science and technology. Natural scientists, engineers, philosophers, historians, and social scientists have by no means arrived at consensus concering the basic character and impact of science and technology. This chapter presents those concepts which enjoy widespread, although not complete, acceptance as adequate descriptions and explanations of these phenomena.

In order to understand science and technology as ongoing human enterprises, it is necessary to grasp their quantitative dimensions. Who is paying for the effect, who is performing the work, and how many people and what types are involved? Chapter 2 is devoted to such questions.

Having outlined these dimensions, the volume then covers the management setting and techniques used in scientific and technological undertakings. Chapter 3 examines the factors unique to the management of research and development and discusses the characteristics which make R&D management distinct from the management of other types of activities.

Next, in Chapter 4, the reader can review some of the promising research frontiers that may help mold this nation's ideas and behavior, including those associated with science and technology themselves.

The last two chapters explain the decisionmaking institutions at the national level having responsibility for science and technology and problems and issues associated with the policies and programs they fashion. Chapter 5 recounts the development of the relationship between science and the federal government in the United States, illustrating the impact of this interface on both science and government. It treats the federal organization of science and technology up to the 1973 changes made by the Nixon administration.

The final chapter summarizes some of the most challenging problems and issues confronting science policy today. It deals with two general classes of problems: (1) those associated with the decisionmaking process associated with fashioning and executing science policies and (2) substantive issues that these policies must address. The first category includes issues such as the role of the science adviser and resource allocation. The second includes issues like the question of economic growth and the depersonalization tendencies of technology.

"The real and legitimate goal of the sciences," Francis Bacon observed, "is the endowment of human life with new inventions and riches."[1] Those charged with the responsibility of developing science and technology as national resources strive to make Bacon's observation a reality. The intellectual raw materials they use, the challenges they face, and the social solutions they initiate, in large measure constitute the subject matter of this volume.

---

[1] Francis Bacon, *Novum Organum*, First Book, in Mortimer J. Adler (ed.). *Great Books of the Western World* (Chicago, Encyclopedia Britannica, 1952), p. 120.

1 The Nature of Science and Technology

Morris C. Leikind and *Wyndham Miles*

Some forty years ago the noted British professor, Alfred North Whitehead, wrote: "The greatest invention of the nineteenth century was the invention of the method of invention."[1] For about a century now man has increasingly mated science to technology, and in the process ushered in a new epoch.

Mankind now learns more in less time. Today we can discover facts that took the ancients centuries to find out. Obviously, this process defies accurate measurement. Tabulating the increase in written material, one scholar estimates that knowledge in the physical sciences has been doubling every fifteen years.[2] At this rate, each generation must absorb more than twice as much knowledge as the preceding generation. This may seem like a Herculean task, but man has gradually adjusted to it. He has been able to accomplish this feat by developing more informed people, greater specialization and cross-fertilization of disciplines. He has, moreover, discarded much obsolete information.

This acceleration has greatly conditioned scientific endeavor whose principal concern is understanding nature, our physical and organic environment, and man himself. By augmenting his understanding of nature, man has created a vast range of science for practical applications, furnishing the basis for technology and contributing to the assets of modern civilization—in transportation, communication, production, and so on.

The Meaning of Science[3]

Science is difficult to define. It represents, on the one hand, a body of objective knowledge about the universe and man's place in it.

---

[1] Alfred North Whitehead, *Science and the Modern World* (New York: Macmillan, 1964), p. 141.

[2] Ellis A. Johnson, "Crisis in Science and Technology," *Operations Research Journal of ORSA*, Vol. VI, No. 1 (January-February, 1958), p. 14.

[3] The following references contain informative treatments of the nature, char-

At the same time, it constitutes an activity, the means by which scientific knowledge is increased. At different times men have held varying notions about the nature of science. Thomas Huxley, famed British scientist, defined it as "organized common sense" [4] and Oliver Wendell Holmes, Sr., medical scientist, essayist, and father of the Supreme Court Justice called it "the topography of ignorance." [5] A more operational definition has been advanced by the American scientist and educator, James B. Conant, who thought that "science emerges from the other progressive activities of man to the extent that new concepts arise from experiments and observations, and new concepts in turn lead to further experiments and observations. . . . The texture of modern science is the interweaving of fruitful concepts." [6] Perhaps a useful, but brief, definition which suggests the dynamic and progressive nature of science might be: Science is cumulative, verifiable, and communicable knowledge.

Science is often subdivided into "pure science" and "applied science." By pure science, often equated with basic research, we mean the investigation of nature to satisfy the need to know. This need to know is independent of whether the discovered facts are useful in any other way except to advance science itself. Applied science, on the other hand, is the application of the results of pure science to some practical human purpose.

The discovery of X-rays by the German physicist, W. C. Roentgen, in 1895, came as a result of studies in pure science. Roentgen was

---

acteristics, and social function of science and technology: Bernard Barber, *Science and the Social Order* (New York: Collier Books, 1962); Jacques Barzun, *Science: The Glorious Entertainment* (New York: Harper and Row, Inc., 1964); Morris R. Cohen and Ernest Nagel, *An Introduction to Logic and the Scientific Method* (New York: Harcourt, Brace, 1934); James B. Conant, *On Understanding Science: An Historical Approach* (New Haven: Yale University Press, 1947); R. J. Forbes and W. J. Dijksterhuis, *A History of Science and Technology* (Baltimore: Penguin Books, 1963); Thomas S. Kuhn, *The Structure of Scientific Revolution* (Chicago: University of Chicago Press, 1962); Robert B. Lindsay, *The Role of Science in Civilization* (New York: Harper and Row, 1963); Derek J. de Solla Price, *Science Since Babylon* (New Haven: Yale University Press, 1961); R. Taton, *Reason and Chance in Scientific Discovery* (New York: Science Editions, 1962); and A. P. Usher, *A History of Mechanical Inventions* (Boston: Beacon Press, 1959).

[4] Thomas H. Huxley, "On the Educational Value of the Natural History Sciences," in his *Collected Essays, 1893* (New York: D. Appleton and Co., 1898).

[5] Oliver Wendell Holmes, in T. Edwards, *Useful Quotations* (New York: Grosset and Dunlap, 1933), p. 569.

[6] Conant, p. 24.

seeking to understand the behavior and nature of cathode rays. It did not take long for persons to recognize that these unknown radiations, called X-rays, by virtue of their ability to penetrate opaque substances and affect a photographic plate, could be applied to medical diagnosis. The discovery of X-rays additionally paved the way for the discovery of radium and radioactivity. These discoveries in turn made possible new insights into the structure of matter and its relation to energy. In 1942 came a highly significant application of the results of years of accumulated basic knowledge through the practical release of nuclear energy.

In 1874 a student in Vienna synthesized a new organic compound as an exercise in "pure chemistry" in order to qualify for his degree. In 1939 an industrial chemist "applied" this compound to a cockroach. The roach died and DDT, a most potent insecticide, found an important application. That discoveries in "pure science" often lead to results of the utmost importance when "applied to practical human use," has become common knowledge.

The Meaning of Technology

Technology is a body of knowledge and devices by which man masters his natural environment. It is not synonymous with applied science, being more empirical in its approach to solving problems. From an historical standpoint, technology arose before the advent of pure science and derives chiefly from common experiences with practical problems. The building of the pyramids of Egypt represents a major technological feat without much recourse to an organized body of scientific theories. The three major technologies—agriculture, medicine, and engineering—all made substantial progress before they were wedded to science. (In fact, it was technology which produced the instruments that helped trigger the scientific outpouring of the 17th century.) The modern world is largely the fruit of the union between science and technology consummated just about a century ago.

Fundamentally, technology is an extension of man's capacity to see further and clearer, hear more and beyond the capabilities of the human ear, travel faster and for greater distances, lift and lower heavier weights with greater ease, and the like. Consquently, technology, unlike some sciences, always contains a human element. In designing a bridge, for example, a civil engineer begins with an actual and potential traffic count which is determined, in large part, by human motivations and socioeconomic considerations.

Thus we see that technology serves to fulfill human needs. Some scholars go so far as to assert that man's unique feature is his capacity to use tools. Yet, closer inspection reveals that the ability to employ tools is shared to a small degree by lesser animals—the female burrowing wasp uses pebbles to scratch over the ground in which she lays eggs; the Darwin finch extracts insects from trees by means of the spine of a cactus. Nonetheless, no other animal has developed tools to the level that man has.

Man has created what we may designate as two general categories of technology: static and dynamic. Static technology includes instruments which involve little or no motion. These instruments are associated with architecture, city building, and hydraulic projects such as dams, reservoirs, and canals. Static technology largely characterized antiquity. Conversely, dynamic technology is based upon power-driven machines (energy converters) such as water mills, sailing vessels, steam and internal combustion engines. The shift in emphasis from static to dynamic technology underlies the meaning of the Industrial Revolution.

Why do men want technology? Practical utilization, discussed heretofore, constitutes only one motivation. At least as important are esthetic elaboration and amusement. Men in the Middle Ages built Gothic cathedrals as much to express the beauty of God as they saw him as to provide a shelter for prayer. In Biblical days, Solomon built his temple for similar reasons. Today, the Golden Gate Bridge provides esthetic pleasure as well as a transportation link across San Francisco Bay.

Technology can provide mass means of amusement that appeal to people in emerging societies who experience great boredom. It is little wonder that such people use their small income first to purchase radios or see motion pictures before they buy more implements to produce economic gain.

Historical Development

One can divide the history of science and technology into several relatively distinct periods. In the preliterate stage, man began to employ technology, including agriculture and medicine; discovered fire; domesticated plants and animals; and fashioned his early primitive tools. The next period, which many call antiquity, was associated with the rise of cities. Societies invented writing and constructed marvelous engineering feats, e.g., the pyramids, Colossus of Rhodes, Roman roads and aqueducts. During this period, Greece, in developing abstract and rational thought, laid the foundation for Western science.

During the Middle Ages, Europe witnessed the decline of classical antiquity and the rise of Christianity. It was left to the Arabs to carry on the pursuit of science. Yet, during the latter years of this period, technological advances occurred, especially within the monasteries; universities arose; Europe saw the beginnings of power technology; and craftsmen invented the clock and spectacles. This progress helped usher in the Renaissance, during which Gutenberg invented printing and, with the work of Copernicus, Leonardo da Vinci, Vesalius, and Agricola, modern science began to stir.

Science flourished during the 17th century. Illustrious scientists (e.g., Boyle, Galileo, Gilbert, Newton, and Harvey) truly made that century a golden age of scientific progress. The organization of science began to take shape with the establishment of scientific societies and journals. At the same time, great strides were made in developing scientific equipment such as the telescope, microscope, and barometer. The latter part of the next century saw the beginnings of the Industrial Revolution, based largely on the steam engine and its application, especially in railways and textile plants.

The following two centuries witnessed a phenomenal increase in scientific and technological knowledge and the impact of this knowledge upon society. In the 19th century scientists achieved momentous advances—in thermodynamics, theory of evolution, cell theory, germ theory of disease, periodic law and electromagnetic theory. In the present century they have produced scientific and technological landmarks in such areas as radioactivity, atomic theory, release of nuclear energy, biochemistry, mass industrial production, aviation and space flight, and in electronics, computers, automation, and cybernetics. Scientific advances in these and other fields have made a profound impact on modern society.

Operating Philosophies and Principles

A scientist generally carries on a project by observation, hypothesis, deduction, and verification—commonly referred to as the scientific method. He may begin by planning experiments which he hopes will give him useful information; he carries them out, observes what goes on, and collects data. When he reaches a certain point in his work, dictated by experience, intelligence and perhaps intuition, he frames a hypothesis, a provisional theory or supposition that explains the facts of a phenomenon. Using this hypothesis as a base, he can deduce consequences. He carries out further experiments to see if the deductions are correct. If they are correct, the hypothesis is confirmed and may become a base for further experiments. If they are not correct, the hypothesis is dropped and another one postulated.

To illustrate how scientists proceed by the above steps, we may consider the periodic classification of the elements. In the early 1800s chemists observed relationships between certain elements and began to suspect that some sort of general relationship existed among all the elements. Attempts to formulate a hypothesis to relate the elements failed; and it might be stated, looking backward, that there were too few elements known to allow scientists to make a good hypothesis. It was as though scientists were trying to guess what a 100-piece jigsaw puzzle looked like with only 25 pieces on the table. As time passed additional elements were isolated and more data were observed. By 1869 enough data were at hand for chemists to come up with a hypothesis that was verified in a number of ways and is still being used today.

Empirical methods based on experience rather than theory are generally resorted to when a scientist enters a new area where he does not have information accumulated by others and where he does not know what to expect. He is guided by his intuition, his curiosity, or his experience. The scientist may be thought of as an explorer. Some explorers bring back much inconsequential information; others bring back highly important facts. Similarly, some empirical scientists discover things of little significance and others things of great moment.

Generally speaking, empirical observation goes on whenever any new subject is attacked, and it continues until sufficient data are accumulated to provide a foundation for future logical work. Two illustrations highlight this proposition. Anton Van Leeuwenhoek (1632-1723), a Dutchman, became interested in optics, developed his own microscope, advanced empirically into microbiology and discovered bacteria, unicellular organisms, sperm cells, blood cells, and other things. Joseph Priestley (1733-1804), a British-American, became interested in science, advanced empirically into the field of gases and discovered oxygen.

Classification and description also comprise an important part of scientific methodology. In any science it is necessary to describe each subject precisely, and thus eliminate confusion when scientists discuss the subject. In many cases, everyday written and spoken language lacks sufficient precision for scentific communication. Furthermore, science deals with things foreign to everyday conversation. For these reasons each science has evolved its own specialized vocabulary. Chemistry, physics, botany, meteorology, and the other sciences have manufactured words having specific meanings and designed to curtail misunderstandings.

Scientists also need a method of classifying things and ideas in each science. A systematic classification shows relationships between ideas and things, aids in describing them, and assists memory. Classifications come about as a result of observation and theoretical speculation. In turn, they have stimulated observations and speculations designed to confirm and expand the classification. A classification has to be open-ended, so to speak, in order that new information may be added or inserted as the science expands.

While classification and description of subjects is common to all sciences, it is particularly important in certain sciences such as botany, zoology, paleontology, and chemistry. In botany, for example, at the beginning (more than 2,000 years ago), there was no systematic classification of all the ferns, grasses, mosses, trees, etc., that make up the vegetable kingdom. Botany made little progress until the 1700s when a Swede named Linnaeus systematized classification within the science, thereby laying its present foundation.

Another technique, tinkering, is more closely allied with technology than with science. A tinkerer is a person adept in one of the crafts, armed with an idea for an invention, proceeding by intuition, and utilizing trial and error methods. Americans of earlier days, perhaps because of their living conditions and philosophy, became particularly skilled at tinkering. The Patent Office files contain tens of thousands of inventions, some very important, that came into existence through tinkering. Among the important tinkerers were Eli Whitney (cotton gin, 1793-1794), John Fitch (steamboat, 1780s), Samuel F. B. Morse (telegraph, 1830s-1840s), Cyrus McCormick (reaper, 1833-1834), Charles Goodyear (vulcanized rubber, 1844), and Elias Howe (sewing machine, 1846).

Some people believe that the day of the individual tinkerer as the chief technological innovator seems to have passed. They contend that a tinkerer finds it ever more difficult to compete with assembled manpower, experience, and facilities of modern industrial firms. Yet, individual tinkerers continue to produce a variety of useful, if not spectacular, inventions.

Fields of Science

Sciences may be sorted into the following groups: (a) mathematics (which can be considered either a science itself or no science at all), (b) the natural sciences, and (c) the social sciences (which, in the opinion of some, are not sciences in the real sense of the term). The natural sciences, in turn, may be sorted into two groups, i.e., physical and biological, and each of these into a number of individual sciences.

Mathematics, a branch of learning all by itself, is concerned with logical arrangement of such things as numbers, quantities and forms. It does not have to be associated with a science or with the external world. A person can practice mathematics without being concerned with the applications to physics, chemistry or any other science. Yet, mathematics in a sense is the language of science. Few sciences, except perhaps descriptive sciences such as botany and zoology, can advance far without the use of mathematics. Mathematics is necessary in expressing the laws of a science, the quantitative relations within a science, and the applications of a science in everyday life.

Some of the most important of the physical sciences are physics, chemistry, astronomy, geology, mineralogy, and meteorology. These sciences deal with the properties of matter and energy encountered in the earth, on the earth, and in space. It is extremely difficult to delineate the physical sciences with precision, but a few general differences stand out.

Both theory and experimentation play important roles in physics and chemistry. Most investigators in these sciences, however, proceed by carrying out experiments and controlling the experiments at will. Physics and chemistry, furthermore, differ from astronomy, geology, mineralogy, and meteorology in that they are concerned with matter and energy throughout the universe and are, therefore, sciences that impinge on or pervade other sciences.

Astronomy, geology, meteorology, and mineralogy may be said to be chiefly observational, rather than experimental, sciences, because the phenomena that take place in or on the earth or in space cannot be repeated at will by the scientist, but can only be observed and described.

The biological sciences are concerned with the process of life and include such disciplines as botany, zoology, and bacteriology. These sciences depend heavily upon classification, observation, and experimentation. They are not as susceptible to measurement as the physical sciences, but are becoming increasingly so.

The various sciences represent artificial categories invented for convenience of study. Often, two or more must be combined to study nature. A chemist, for example, who studies petroleum is a petrochemist; one who studies living matter is a biochemist. Because we now need to know more about the role of fats and oils in the body, petrochemists and biochemists have to be concerned with each other's fields. We now have biophysicists, physical chemists, neurochemists, astrophysicists, astrobiologists, geochemists and geophysicists, phytochemists and zoogeographers as well as medical entomologists. As techniques and ideas fruitful in one field prove applicable in other branches of science, more and more researchers acquire interdisciplinary skills. For example, increasing numbers of engineers are entering medical research, designing electronic equipment to further the science of physiology.

Creativity

While scientists and inventors may apply sound operating philosophies and principles, not all are creative. Many minds pondered about the nature of the universe, but relatively few, such as Newton and Einstein, have shaped our fundamental understanding of it. While we can't define creativity with any precision, or know exactly how it works, we do know that it constitutes the single most important force contributing to scientific and technological progress.

Over the years, researchers have increasingly investigated the nature of creativity. While no inviolate hypothesis has emerged, men have come to understand this phenomenon better. Creativity is the ability to produce something new. Working with known facts, the creative mind by some leap of imagination develops novel combinations. One can form such combinations by perceiving unexpected and significant relationships and distinctions in existing knowledge. Newness can take several forms—new solutions, new ideas, new uses for old things. For example, throughout the ages men have found ever-increasing applications for the wheel. At first, it was largely confined to use on land vehicles, chiefly wagons and chariots. Someone thought of attaching buckets, partly submerging them in water, and tapping the resultant power generation—and the water wheel was born. Later, someone turned the wheel on its side, attached guide vanes, thereby creating a turbine. Some put a rope around it (the pulley), notched it (the gear), and raised it up to harness wind (the windmill).

The making of new combinations is the creative process. Points of view concerning the creative process vary. One view, held by but a few, asserts that creativity is purely accidental and owes nothing to reason or logic. A plausible view holds that chance is important in observational or experimental sciences; but, as Louis Pasteur once aptly observed, chance helps those minds which are prepared for it. Scholars and researchers still debate the relative effectiveness of the logical process, characterized by sustained and methodical effort, as opposed to intuition (hunches). Rather than argue this question, it appears more profitable to examine the creative process by analyzing certain stages which some researchers have found to be present: concentration, incubation, illumination, and verification. This approach enjoys the additional merit of placing both logic and intuition into a general scheme.

In the first stage occurs the stirring of ideas; the mind undergoes a conscious, but fruitless effort to find significant relationships and distinctions. Some claim that this is a period of intellectual chaos. During the incubation period, although the mind appears at rest, the subconscious processes ideas. It is during the period of illumination that the mind achieves the sought-after insight. This stage is often described as a blinding flash of creative thought unleashed by intuition and many people mistake it for the entire creative process. The last stage, verification, involves testing the new idea against reality. Judgment is called into play as the individual seeks to prove the validity of his ideas. These stages interweave and a person need not follow them progressively nor exactly in sequence.

We also know that creativity is not entirely an intellectual effort. More often than not, creative people exhibit an emotional commitment to their ideas and products. The act of creativity becomes highly personal. For this reason, science has not flourished in ages stressing anonymity, such as in the age of medieval crafts. The creative man usually craves recognition for his work; he seeks a personal identification with his product. In addition, he seeks the satisfaction of seeing his handiwork become part of the intellectual landscape upon which future scientists and inventors will tread. Lastly, creative scientific and technological works gratify man's esthetic sense. Newton saw in the laws of gravity, and Edison in the electric bulb, somewhat the same beauty that Da Vinci saw in the Mona Lisa or Shakespeare saw in the soliloquies of Hamlet.

Relationship to Resources

In their modern context, science and technology more than ever before should be considered as national resources, both in themselves and as tools for improving other material and human resources. They constitute resources because they provide sources of supply and support that man can use. The technical knowledge that he brings to the task of constructing an airplane is as much a resource as are physical materials, such as aluminum or titanium. Science and technology represent assets that, like other assets, can be augmented or conceivably depleted (if man forgets knowledge as he did during the early Middle Ages). As in natural resources, their more promising "veins" can be tapped, existing knowledge can be processed into new "intellectual products"; and utility and cost determine the importance of their social and economic applications.

Science and technology increase the value of material resources by amplifying their usefulness. While today the world may enjoy substantial reserves of many raw materials, due in considerable degree to the application of science and technology, future demands may bring

difficult resource problems. Some areas will feel the pinch on exploitable resources sooner than others. It is well known, for instance, that
the United States has reverted in some measure from a materials-
surplus to a materials-deficient nation. In large part, the chief resource
problem of the United States is not a total lack of certain natural
materials, but a dearth of these materials in a form which can be
exploited economically. Since this country will not knowingly let itself
become a resource pauper, it must not only discover new deposits, but
also find ways—scientific and technological—to exploit more prudently
its indigenous natural resources.

No doubt our most important resource, one which utilizes all others,
is man. While the material sectors of our society have experienced
tremendous advances, the human being still remains their initiator and
user. Man and his science and technology are mutually dependent in
several ways. First, only a healthy and educated population can produce the scientists and technologists who can promote the growth of
science and technology. On the other hand, science and technology
are needed to develop such a population. Second, man must design
his social institutions in ways compatible with science and technology.
Man's ideas and institutions can conflict with his scientific and technological products and procedures. Consequently, the effectiveness of
human resources both affects and is affected by these two activities.
Lastly, the uses to which human resources are put, depend in part
upon science and technology. For example, as science and technology
improved agricultural productivity, the size of our farm population
fell from 40 percent at the turn of the century to 8 percent in 1965.

Effects on Economic Strength

Economists agree that a positive correlation exists between progress
in science and technology and economic growth; they differ on how far
this correlation extends. Some argue that economic growth stems from
more than the results of research and development. Some prefer to
emphasize the many factors other than technological innovation—such
as economies of scale, a more educated citizenry, shifts in population,
effects of mineral exploration and findings—which stimulate increased
production. Yet, in modern times, large-scale scientific and technological programs represent more than just another factor contributing
to economic growth; they represent a crucial element.

Today, economic growth depends upon the economy's increased capability for production and distribution. At heart it means growth in a nation's ideas and skills. Further, investment provides the elixir of economic growth. While economic factors determine the investment rate, new ideas, new developments in science and technology, innovations such as new products, new processes, new resources, and the like, provide the springboard.

The great discovery of our age is that technological innovation need not be haphazard. Industry and government have developed a new concept of planned and systematized innovation, founded on vastly expanded scientific and engineering efforts. These institutions are now making regular provision for the occurrence of new and unpredictable developments. In fact, "the discovery of systematized innovation may turn out to be a qualitative change in the economy—one having the same importance for future growth as the development of the concept of capital investment itself had during the past two centuries." [7]

Past Influences

A few selected examples of past effects of scientific investigations and technological applications upon society illustrate how much these activities have changed the world. Technological achievements have vastly accelerated the speed of human travel. Until horses were domesticated in the late Neolithic or early Bronze Age, about 15,000 B.C., man's fastest speed of travel was on foot, at a top speed of 10 miles per hour. From the domestication of the horse to the invention of the locomotive, about 1825, man increased his maximum speed to some 35 miles per hour. One hundred years ago circumnavigation of the globe in 80 days was considered remarkable. Aircraft now travel faster than sound and astronauts in space capsules orbit the earth in 90 minutes.

The development of engines has unleashed potent forces in revolutionizing the modern world. The steam engine developed in the 18th century helped trigger the Industrial Revolution. The internal combustion engine, fashioned less than a century ago, wrought other profound social changes. The mass production of the automobile put

---

[7] "The U.S. Invents a New Way To Grow," *Business Week* (January 23, 1960). See also Leonard Silk, *The Research Revolution* (New York: McGraw-Hill, 1960).

this nation and much of the world on wheels. It led to vast networks of highways that lace this country together from coast to coast. It was this same type of engine which permitted man to free himself from the shackles of gravity, enabling him to conquer the air. The rocket engine now allows him to explore outer space.

The invention of firearms revolutionized warfare, and the impact of this technological development has not spent itself on our society. The need for metals for cannons and hand guns stimulated mining and metallurgy. The need for accurate methods of aiming cannons gave rise to the science of ballistics. Theoretical studies on the movement of projectiles contributed to our knowledge of the dynamics of falling bodies and to our understanding of the law of gravity. The internal combustion engine was another byproduct of the invention of the firearm. This engine is nothing more than a cannon with a piston in it, attached to a crankshaft, actuated by the explosive force of petroleum vapors.

Atomic energy, which made its advent in the closing days of World War II, has already had far-reaching consequences possibly exceeding that of the invention of gunpowder. Man now has within his grasp literally limitless energy tools to destroy or to build. The dangers are well known and condition much of modern international life. Yet, we not only have within our hands the potential of nuclear energy to power much of our industrial plant, we also use this type of energy for propulsion—notably for ships, submarines and aircraft carriers. There is also considerable confidence that in the foreseeable future nuclear power may propel vehicles probing deep space.

Summary

Science and technology can be viewed both as intellectual activities and as resources. Man has always considered technology as a resource. Today, science has added social utility to its original function of expansion of knowledge. The methodologies of science are fashioned to permit the scientist to come as close as possible to the truth of the universe, while those of technology respond more to human needs. In both, the driving force is creative effort that, more than ever, provides a vital ingredient for economic strength and social progress.

2 The Dimensions of Research and Development[1]

Ralph Sanders

Research and development (R&D) continue to occupy a central position in the nation's complex of science and technology. In the most general sense, the new or improved knowledge, understanding, products, and processes created by research and development enable us to sustain a dynamic economy and a strong defense posture. Portions of the vast stream of scientific information which flow from and feed back into R&D are eventually codified, popularized, and disseminated throughout the nation for the training and education of persons in industrial production, medical care, government services, and the like. The nation's scientists, moreover, receive their fundamental training through the performance of research in university laboratories, under the tutelage of experienced academic scientists.

"Research and development" is a composite term that covers a wide gamut of activities in the natural and social sciences, ranging from the most esoteric and abstract mathematical theory to the most practical design, test, and evaluation of a new device. The nature of activities generally subsumed under research should be clearly distinguished from those making up development. The social sciences, such as economics, sociology, and anthropology, constitute a relatively small portion of the disciplines considered within the category "research and development."

Research embodies a quest for knowledge and understanding, a pushing back of the unknown. Whether the understanding is sought primarily with the cultural goal of enlarging man's comprehension of the universe or with an immediate practical application in view, research is concerned with the "how" and "why" of natural and social phenomena.

Within this generalization, the distinction between *basic research* and *applied research* is often made principally in terms of motivation.

---

[1] This chapter is based on information and statistical data provided by the National Science Foundation.

Basic research is conceived of as primarily concerned with achieving fuller knowledge or understanding for its own sake, while applied research has a practical objective for obtaining the understanding. The concept of basic research embraces both "pure" research undertaken solely to enlarge our understanding of natural or social phenomena and "mission" or "goal-oriented" basic research. The latter is undertaken in the hope or expectation that the understanding gained will eventually contribute to the applied research which is geared to the operating purposes of the supporting organization.

In the federal government, it is the mission of the National Science Foundation to be the prime supporter of non-goal-oriented basic research. Outside the federal government, state and local governments, colleges and universities, and certain other nonprofit organizations, particularly philanthropic foundations, support this type of scientific inquiry. Other important basic research programs conducted by the Department of Defense, the National Aeronautics and Space Administration, the Department of Health, Education and Welfare, the Atomic Energy Commission, and a number of industrial corporations are related in a general way to various aspects of major government operating responsibilities or company commercial goals. Although no official estimates have been made, it is clear that mission-related basic research accounts for a much greater volume of funds than does pure basic research. The preference for mission-oriented research has become stronger recently; in the past few years even the National Science Foundation has placed increased emphasis on stimulating applied research to help solve domestic problems.

Development, on the other hand, draws on the results of research for the design and production of new or improved products and processes to meet well-defined needs. More specifically, it is the systematic application of scientific and technological knowledge to the production of useful materials, devices, systems, methods, and the design and development of prototypes and processes. A further distinction must be made between development and production. If the primary objective is to make further improvements on the product or process, then the work comes within the definition of R&D. If the product or process is substantially "set," and the primary objective is to develop markets, do preproduction planning, or initiate a smooth production process, the work is no longer R&D.

The scope and trends of R&D in the United States are best indicated by statistical data, chiefly those dealing with financial and personnel resources, and compiled by various government agencies, principally the National Science Foundation. A wide range of goals and stimuli underlie the statistics of R&D expenditures. In particular, outlays for applied research and development are usually associated with larger and sometimes overlapping activities such as industrial production, national defense, improvement of health, pollution abatement, mass transport, and the conservation and exploitation of natural resources. Judgments on whether the total amounts for applied research and development are too high or too low or rising too fast or too slowly cannot be made apart from a consideration of progress toward the goals with which the component portions of R&D are associated.

Data on R&D funds or personnel cannot be equated with the input of creative energy or the output of new knowledge, products, and processes. Furthermore, the dollar data do not represent what *should* be spent or what *can* be spent but simply what *is* currently expended, in dollars, in the conduct of R&D. Because of the difficulties of making fine distinctions between component activities, data on the larger aggregates such as total R&D are recognized as being stronger than those for smaller breakouts such as basic research.

General Dimensions[2]

R&D in the United States has grown enormously since 1940. World War II launched R&D as a large-scale national activity. Investment in R&D gathered momentum during the 1950s, and showed a compound

---

[2] The National Science Foundation is the major compiler and publisher of data on R&D in the United States. Statistics pertaining to financial aspects of R&D can be found in the following NSF documents: *National Patterns of R&D Resources: Funds and Manpower in the United States—1953-1973* (Washington, D.C.: National Science Foundation, February 1973) (NSF 73-303); *Federal Funds for Research, Development and Other Scientific Activities: Fiscal Years 1971, 1972, 1973,* Survey of Science Resources Series, Vol. 21 (Washington, D.C.: National Science Foundation, August 1972) (NSF 72-317); *An Analysis of Federal R&D Funding by Budget Function: 1969-1974,* Survey of Science Series, 1973 (Washington, D.C.: National Science Foundation, October 1973) (NSF 73-316). See also U.S. Office of Management and Budget, "Federal Research and Development Programs," *Special Analyses of the United States Government: Fiscal Year 1975* (Washington, D.C.: U.S. Government Printing Office, 1974).

rate of growth of 17 percent for the decade. By mid-1960 the United States was pouring some $14 billion a year into R&D. Because of price level increases, however, the percentage increase in real terms was less than that reflected in the dollar figures, but still very appreciable.

Since the mid-1960s, support for R&D has undergone marked change. Decreased emphasis on federally supported defense and space R&D has been more than offset by increases in nonfederal R&D. While the economic slowdown at the end of the 1960s caused a leveling in non-federal R&D funds, total R&D expenditures reached an estimated $30 billion in 1973. This R&D investment amounted to some 2.4 percent of the nation's gross national product, down from 2.5 percent in 1972 and from 3 percent in 1964. The present Administration increased stress on applying R&D to help alleviate some of the country's major problems, a step that would seem to mean little, if any, further drop in this percentage, and perhaps even an increase.

Among the various private organizations concerned with R&D, phil-anthropic foundations are the chief providers of research funds. Col-leges and universities are essentially performers. They generally con-centrate on research, and receive a major portion of financing from the outside. Most federal agencies combine the two functions. Private health and welfare organizations act chiefly as intermediaries and col-lect funds from many sources for distribution to research organizations.

Figure 1 shows that the largest source of R&D funds is the federal government; it accounted for some 53 percent of the $30 billion total in 1973. By contrast, in 1964 the federal government contributed 66 percent. Except for a few public laboratories, the federal government supported practically no research a little over 30 years ago. Industry finances two-fifths (41 percent) of the total R&D in the country, and colleges and nonprofit institutions support the remainder, about 6 percent.

If R&D in the United States is heavily financed by public means, it is the private institution that does most of the work—some 85 percent of the total (see Figure 1). This reflects the strong confidence that Americans place on the productivity and reliability of their private sector. Thus, nongovernmental organizations, including industry, uni-versities, and nonprofit institutions, carry on the bulk of scientific and engineering investigations. Industry alone did over two-thirds (67 percent) of all R&D in the country in 1973. In providing funds in

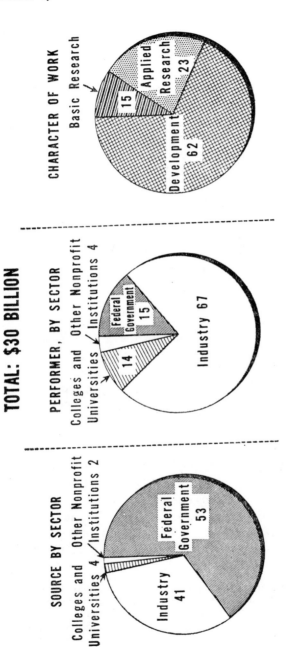

TOTAL: $30 BILLION

CHARACTER OF WORK

Basic Research 15
Applied Research 23
Development 62

PERFORMER, BY SECTOR

Colleges and Universities 14
Other Nonprofit Institutions 4
Federal Government 15
Industry 67

SOURCE BY SECTOR

Colleges and Universities 4
Other Nonprofit Institutions 2
Federal Government 53
Industry 41

Source: National Science Foundation

Figure 1. Percent Distribution of Research and Development Expenditures, 1973 (preliminary).

support of research, the federal government enables private institutions to carry on R&D far beyond their internal financial capabilities.

Figure 1 highlights another important characteristic of R&D, namely, that the development side of R&D receives more than 1.5 times the support accorded research. The reasons for this are easy to find. First, development offers more immediate rewards and men tend to rationalize resource use in utilitarian terms. The fruits of applied, and especially, basic research may not be gained for years. Generally speaking, it is also more expensive to fashion hardware and industrial processes than to discover facts of nature or construct scientific theories. While it is true that scientists have long ceased to rely on "pencils, paper, and sweat" and the price of some scientific equipment has sky-rocketed (e.g., a new nuclear accelerator is estimated to cost $300 million), the cost of fashioning certain engineering models may require even greater resources. The prototype of an advanced ballistics missile costs much more than the pioneering research into the DNA (deoxy-ribonucleonic acid) molecule in biology.

Even within research, the applied aspects receive greater support than basic investigations. Yet, despite the heavier emphasis on the applied side, expenditures for basic research have increased by about $4 billion, from $489 million in 1953 to some $4.5 billion in 1972. During the decade 1963 to 1973, basic research expenditures registered a gain of over 100 percent in current dollars and nearly 50 percent in constant dollars (less inflation).

Industrial R&D

Industrial spending for R&D reached a level of $18.3 billion in 1971. During the 1960s, industrial R&D funds increased at the rate of 7 percent a year.[3] In constant dollars, this increase has been 4 percent annually. However, in real terms a decline has taken place in industrial R&D spending. While the 1971 figure represents a 1 percent increase over 1970, in real terms a 3 percent decrease took

---

[3] This discussion of distribution of R&D by industry and firms is based upon National Science Foundation, *Research and Development in Industry, 1971 Funds, 1971 Scientists and Engineers* (Washington, D.C.: National Science Foundation, January 1972) (NSF 73-305) and National Science Foundation, "Industrial R&D Spending, 1970," *Science Resources Studies Highlights,* December 10, 1971 (Washington, D.C.: National Science Foundation, 1971) (NSF 71-39).

place. The decline in federal funds to industrial R&D chiefly accounted for the decrease.

At the turn of the past decade, industrial funding for basic research and development took a slight downward dip. This slide halted in 1971 when basic research in industry consumed almost $625 million (about 3 percent of total R&D expenditures). Applied research used $3.4 billion (18 percent). Development, which accounts for 79 percent of industrial R&D spending, amounted to $14.3 billion in 1971. As these figures indicate, industry is committed chiefly to producing useful products from its R&D. Even its basic research has as its objective an application, however remote, or an explanation of an unknown phenomenon that appears in its development activities. This type of basic research is often referred to as "biased basic."

R&D activity has been concentrated in a few industries. In 1971 over one-half (52 percent) of industrial R&D funds were spent by two industries—aerospace (aircraft and missiles) and electrical equipment and communications. Looked at another way, five industries spent over 81 percent of the total. In addition to aerospace and electrical-communications, these five included chemicals and allied products, machinery, plus motor vehicles and other transportation equipment. On the other hand, industries such as textiles and apparel, lumber and wood products, as well as paper products have supported very modest R&D efforts.

The federal government was the source of 42 percent of the R&D dollars spent by industry in 1971, down from 58 percent in 1963. Only the aircraft and missiles and electric equipment industries derived the major share of their support from federal sources in 1971— some 80 percent and 50 percent respectively. Since 1958, the five leading industries increased their company-financed research support at a much faster rate than federally financed; the most dramatic changes occurred in the electric equipment and machinery industries.

Large-sized companies carry on most major R&D activities. In 1971, companies with 10,000 or more employees conducted over four-fifths of all industrial R&D. Many of these companies were in the five industries in which R&D efforts are concentrated. In the non-manufacturing industries where most R&D is conducted by small firms, only 5 percent was undertaken by companies with 10,000 or more employees.

Federal Contributions

The growth of federal support for R&D (including R&D plant expenditures) has been remarkable. In 1973 the federal government spent an estimated $16.8 billion as compared to $74 million in 1940. The inflationary factor decreases the extent of this growth in terms of R&D purchasing power, but is counterbalanced to some degree by the increase in the productivity of R&D. Even discounting inflationary effects, the differences are surely impressive. As a percentage of total federal spending, R&D expanded from 0.8 percent in 1940 to about 15 percent in 1967, an eighteenfold increase. However, since 1967, the ratio of the R&D budget to the total federal budget has decreased each year. In both 1972 and 1973 it fell to 7.4 percent, the lowest in the entire 1960-1973 period.

In the early 1970s, four federal agencies received the lion's share of R&D monies—the Department of Defense (DOD), the National Aeronautics and Space Administration (NASA), the Department of Health, Education and Welfare (HEW), and the Atomic Energy Commission (AEC). These are listed in descending order of magnitude. Of these, DOD, NASA, and the AEC, all of which are involved in national security, consumed the largest portion, over 75 percent of all federal R&D spending, down from 90 percent in the mid-1960s. While the DOD share remains high, it dropped from 80 percent in 1955 to about 50 percent in 1973, accounting for over $8 billion. From NASA's establishment in 1958 to the mid-1960s, its funds grew at a rate of 57 percent a year, compared to less than 10 percent for DOD. In 1964, NASA spent some 34 percent of every federal R&D dollar. Since that year, NASA's share has dropped sharply, and in 1973 it accounted for somewhat over 20 percent.

HEW's estimated federal obligations in 1973 of $1.8 billion accounted for a little more than 10 percent of the total, up from 7 percent in 1967. While HEW received less support than the AEC in the mid-1960s, in 1973 it obtained some $480 million more and its relative funding is expected to grow even more in the near future. The overwhelming portion of these funds supports the Public Health Service which includes the highly respected National Institutes of Health.

Agencies receiving the next highest funding received less than one-third that of the AEC. The National Science Foundation (NSF) obligated about $480 million in 1972, less than 2 percent of the federal total. These monies, however, are crucial in supporting non-mission-oriented projects which other federal agencies often do not underwrite. Of interest is the fact that in recent years the Department of Transportation has surpassed the Department of Agriculture and other agencies as a recipient of federal R&D funds. This shift, no doubt, reflects the increased attention being paid to solving problems of mass transportation.

Between 1967 and 1973 the distribution of federal funds among major R&D activities has shifted somewhat. The federal government continues to emphasize development, but this activity has fallen from 66 percent to somewhat less than 60 percent of the total. Basic research funding, on the other hand, rose from 12 percent to 15 percent while monies for applied research increased from 20 to 26 percent. The chief retarding effect on development funding has been the curtailment of NASA programs, but in some years from reductions in defense programs.

Over the past few years the relative shares of the life sciences, the environmental sciences, and the social sciences have tended to rise while the shares of physical sciences and engineering have tended to fall. Psychology and mathematics have remained stable. In 1973 the engineering sciences accounted for an estimated 31 percent of the federal research total; the life sciences 29 percent, the physical sciences 17 percent, the environmental sciences 11 percent, and the remainder about 12 percent. Despite increased funding for the social sciences, they still accounted for only 6 percent in 1973.

The DOD allocates a large portion of its money to development of major weapon systems and most of its research money to the physical sciences. NASA also spends most of its money in developing hardware for space exploration. Increases in programs, such as the Skylab and the space shuttle system, and satellites for understanding and managing earth resources almost offset decreases in the Apollo lunar landing program. Similarly, while the AEC cut back its development efforts in underground testing of nuclear weapons, space propulsion, and space electric power, it accelerated the development of an economic liquid-metal, fast-breeder, power reactor. HEW increased support of innovative educational programs, heart disease and cancer research.

The NSF inceased its financial support, in part, by assuming responsibility for programs previously conducted by other agencies, such as the National Magnet Laboratory from the Air Force and the interdisciplinary materials sciences laboratories from the Defense Advanced Research Projects Agency. The new Environmental Protection Agency also has picked up R&D programs formerly performed by the Departments of HEW, Interior, and Agriculture. These R&D efforts in the fields of water and air pollution, pesticides, radiation, and solid wastes could very well receive increased stress in the future.

Manpower[4]

One vital ingredient (some contend it is perhaps *the* vital ingredient) of a national capability in science and technology is manpower—chiefly scientists and engineers. Manpower, rather than money or materials, is sometimes the pacing item in many research and development projects. Vannevar Bush, who headed the nation's top scientific organization during World War II, enthusiastically supported the assertion of his colleague, James B. Conant, that "in every section of the entire area where the word science may properly be applied, the limiting factor is a human one. We shall have rapid or slow advance in this direction or that depending on the number of really first-class men who are engaged in the work in question."[5]

While manpower statistics are necessary in discussing the dimensions of R&D, it should be recognized that manpower analysis lacks the precision of the hard sciences. Forecasts and projections often are difficult and invariably must be hedged with qualifications. Two chief reasons account for the difficulty of analyzing the work of scientists and engineers. First, it is almost impossible to measure the professional output of an individual in an objective way. Salary, job description, and educational background are not adequate determinants of a man's contribution. The character of management, the work environment, and the intellectual challenge all influence creative productivity. Second, there are variations in basic job tasks, from purely technical jobs to purely managerial tasks and from professional

---

[4] Manpower statistics can be found in National Science Foundation, *National Patterns of R&D Resources, 1953-1972* (Washington, D.C.: National Science Foundation, December 1971) (NSF 72-300).

[5] Quoted in Vannevar Bush, *Science: The Endless Frontier* (Washington, D.C.: Public Affairs Press, 1946), p. 59.

to technician status. Thus, analysis of scientific and engineering manpower can only have limited meaning if expressed in numbers alone. On the other hand, the data available do illustrate the profile of our scientific and technical manpower in a useful way.

The nation's science and engineering manpower force has grown considerably since the pre-World War II era. These people serve the nation in many ways; some expand scientific knowledge by doing research; some apply scientific information and engineering techniques to develop new products and services, or to solve problems in health, defense, or transportation; some operate complex systems for communications, services, and space exploration; and some educate and train manpower.

In 1973 the pool of active scientists and engineers engaged in R&D totaled over half a million. Many more of these professionals conducted work other than R&D. The proportion of all natural scientists and engineers maintaining R&D activities grew to a high of 37 percent by 1964, and dipped slightly to one-third by 1973. The absolute numbers of R&D scientists and engineers leaped by more than 110 percent in the 11 years between 1954 and 1965, but dropped sharply to only 5 percent from 1965 to 1971.

Concurrent with the decline in the proportion of scientists and engineers performing R&D in the late 1960s was a major increase in unemployment among these professionals. In mid-1971, for example, 3.5 percent of all R&D scientists and engineers were unemployed, in contrast to 1.9 percent of all other scientists and 2.6 percent of all other engineers. The slowdown in the economy and the reduction of federal contracts in the aerospace industry during this period largely accounted for this higher unemployment. In late 1971, however, unemployment began to decline slowly, improving still more in 1972.

In 1972, industry employed somewhat less than 360,000 R&D scientists and engineers, some 130,000 more than in 1957. From 1957 to 1969 the number of R&D scientists and engineers grew at an average annual rate of 4.5 percent. Conversely, from 1969 to 1972, employment fell by about 8 percent of the 1969 work force. By 1972 employment was down to about the 1966 level. The major reason for this decline was the significant slowdown in activity in the aerospace and communications equipment industries, sparked in large part by a cutback in federal orders. Moreover, since 1963 there has been a decline in the proportion of R&D scientists and engineers in industry supported by federal funds, from 48 percent in 1963 to 34 percent in 1972.

The technology-intensive industries naturally employ the greatest number of R&D scientists and engineers. In 1972 the electric equipment and communication industry employed the single largest bloc of R&D scientists and engineers, about 88,000. Not only was this industry the biggest employer, but between 1957 and 1972 it experienced the fastest growth, more than doubling its use of these professionals. The second highest number, some 73,000, worked for the aircraft and missiles industry. The chemical and machinery industries came next, each tapping the services of about 41,000, followed by the automotive industry with some 30,000, and the scientific instruments industry with 19,000.

Between 1963 and 1972 there has been a declining trend in the proportion of R&D scientists and engineers supported from federal funds in all industries except petroleum refining and extraction. In this industry the share of federally funded R&D employment advanced from 3 percent to 9 percent. In the large electrical equipment and communications industry this percentage dropped from 56 to 43 with the aircraft and missile industry similarly declining from 84 percent to 76 percent.

The number of scientists and engineers at colleges and universities engaged primarily in R&D activities increased from almost 39,000 in 1975 to 47,000 by 1971. Life sciences accounted for by far the largest portion of the total—some 65 percent in 1971. In contrast, all physical scientists (three-fourths of whom are chemists and physicists) represented only 15 percent and mathematicians only 3 percent of the total. Engineers accounted for about 10 percent, while the social sciences and psychology combined amounted to 6 percent.

In 1971, federally funded research and development centers administered by universities (e.g., Livermore Laboratory run by the University of California) employed almost 11,000 scientists and engineers, down slightly from the 11,055 employed in 1965. Engineers and physical scientists totaled 46 percent and 29 percent, respectively, of the total. The high proportion of these professionals resulted from the fact that these centers concentrate on applied research and development with the objective of eventually fashioning useful hardware. Nonprofit institutions employed a larger group of R&D scientists and engineers, over 21,000 in 1970. Life sciences here was the largest field of specialization, accounting for 34 percent. Engineers and social scientists each accounted for 22 percent of the total.

The federal government employed over 46,000 R&D scientists and engineers (exclusive of R&D managers) at the beginning of 1971. Over half, some 52 percent, were at the Department of Defense. The next largest group was the 13 percent at NASA, followed by 10 percent at the Department of Agriculture. Other federal departments employing significant numbers of R&D scientists and engineers include HEW, Interior, and Commerce.

Summary

The R&D effort in this country is heavily financed by the federal government, but is carried on mostly by private institutions. Within the recent past, however, the federal share of total R&D support has fallen. Within government, agencies concerned with national security consume by far the largest slice of the federal R&D dollar, and most of this money goes to development rather than to basic research. Finally, over 2½ million Americans are employed in science and technology. (This total includes not only those in R&D, but also those in education, in science and engineering, in production, and in public planning and regulation.) A little over half of those so employed are engaged in R&D activities and most of these work for the relatively large industrial firms. All in all, a statistical profile of science and technology in the United States illustrates that R&D activities represent a major social and economic force. Any understanding of science and technology as national resources demands an awareness of these statistical realities.

3 Managing Research and Development Activities

Ivan Asay

The benefits to be derived from our vast investments in science and technology depend in great measure upon how effectively we manage research and development.[1] Inherently complex even on a small scale, the management of organized scientific and technological investigations today presents the manager with many perplexing problems, partly because the scale of research and development has grown so large and so fast.

Most observers agree that there are important differences between the management of research and development and the management of other kinds of human activity although they debate the extent and nature of these differences. The uniqueness of management in this area is said to be more marked at the research end of the R&D spectrum. At the other end, as we approach product engineering, the management problems more nearly resemble those of production-oriented organizations. Accordingly, we shall give greater consideration in our

---

[1] The literature on the management of research and development is extensive. The following works contain useful analyses of the subject: George P. Bush and Lowell H. Hattery (eds.), *Scientific Research: Its Administration and Organization* (Washington, D.C.: American University Press, 1950); C. C. Furnas, *Research in Industry* (New York: D. Van Nostrand, 1948); Arthur Gerstenfeld, *Effective Management of Research and Development* (Reading, Mass.: Addison-Wesley, 1970); David B. Hertz, *The Theory and Practice of Industrial Research* (New York: McGraw-Hill, 1950); Maurice Holland (and contributors), *Management's Stake in Research* (New York: Harper and Brothers, 1958); Fremont E. Kast 1970); David B. Hertz, *The Theory and Practice of Industrial Research* (New York: McGraw-Hill, 1963); Charles V. Kidd, *American Universities and Federal Research* (Cambridge, Mass.: The Belknap Press of Harvard University Press, 1959); Robert T. Livingston and Stanley H. Milberg (eds.), *Human Relations in Industrial Research Management* (New York: Columbia University Press, 1957); C. E. Kenneth Mees and John A. Leermakers, *The Organization of Industrial Research* (New York: McGraw-Hill, 1950); Stuart G. Montieth, *R&D Management* (Hartford, Conn.: Daniel Davey, 1970); Jack A. Morton, *Organizing for Innovation: A Systems Approach to Technical Management* (New York: McGraw-Hill, 1971); James B. Quinn, *Yardsticks for Industrial Research* (New York: Ronald Press, 1959); and Daniel D. Roman, *Research and Development Management* (New York: Appleton-Century-Crofts, 1968).

29

discussion initially to the special features of the management of research. We will reserve until later our review of some of the unusual aspects of the management of development and engineering activities.

The distinctive characteristics of research management may be viewed as relating to the intrinsic nature of scientific work, the traits and preferences of dedicated researchers, and the consequent institutional and environmental situation in which scientists can be most creative. Taken together, these considerations have led to a commonly held thought that research can be managed in a businesslike way, but not in a way like other businesses.

The Researcher in the Organizational Setting

The hallmarks of scientific research are originality and creativeness —as opposed to repetition and continuity. Researchers endeavor to extend the known into the unknown with considerable unpredictability of outcome and a high risk of disappointment. The usual management tasks of preparing plans, gauging progress, and measuring or even recognizing results tend to be performed mostly at the bottom of the research organization rather than at the top as in other organized activities. Nonspecialized administrators up the line often have difficulty in appreciating the judgments made by their scientist-researchers and in sympathizing with the ingrained pressures on scientists to pursue research tasks whose significance is understood only by them.

Stereotypes of the scientists include the long-haired recluse, the temperamental prima donna, and the brilliant young prodigy. In reality, scientists and engineers are a varied group of human beings whose image has been colored by a few unusual individuals. Still, scientists as a group do tend to exhibit certain special characteristics. They have above-average intelligence and education and a well-cultivated respect for facts and knowledge. In many cases, they are young and have short work experience compared to their level of responsibilities. Single-minded focus on a chosen specialty, dedication to work, and impatience with diversionary influences are common personality traits.

In order to understand better the blend of order and independence that is required in effective scientific pursuits, it is important to recognize that both science and its cousin, technology, are social products. Although we recognize that scientists are frequently men of high intellectual endowment and academic attainment, we must also recognize that scientific "breakthroughs" often come at more or less predictable times in history after a number of scientists have been working simultaneously on the same problems and have been in communication with each other. Also we note that scientists, like other men, have been subjected to two large-scale social processes that have affected practically all of the modern world of work. We refer here to the process of professionalization and the process of bureaucratization.

Professionalization, Bureaucratization, and Adaptive Mechanisms[2]

Whereas scientists were essentially amateurs at one time (for example, in the days of Benjamin Franklin), they have now become organized into professional groups. These more professionalized occupations are characterized by (1) highly specialized intellectual skills based upon training in a general body of underlying theory, (2) formal occupational associations that control entry and develop codes of ethics to govern the occupationally related behavior of members, and (3) general community recognition of the special occupational status of professional members, often resulting eventually in some form of certification or licensing. Professionals develop rather strict standards of behavior, but their behavior is commonly controlled by mechanisms that are deeply impressed into the personalities of professional people.

In a work situation, professionals, trained to respond primarily to internal standards of control, tend to resist external controls from those outside their profession. They are likely to seek jobs that allow a high degree of individual professional expression and opportunity to move from one organization to another. Also, persons trained in different professions tend to differ in their values and standards. For example, physiologists are prone to take great pride in making contributions to scientific knowledge and prefer a university career while mechanical engineers prefer solving technical problems and a career in industry. There are problems in employing scientists in development activities and engineers in research activities. Research differs from development not only in objectives, but also in the kind of professionals who are best suited to attaining these objectives.

Much of the conflict between scientists and managers stems from the fact that scientists today work in industry and government in increasing

---

[2] The discussion on professionalization and bureaucratization and adaptive mechanisms of both scientists and organizations was prepared by Dr. Howard M. Vollmer.

numbers and are subjected to the process of bureaucratization. As originally described by the German social scientist, Max Weber, bureaucratization involves a formalization of the structure of administration. This process brings about a clearly defined hierarchy of authority which appears as a pyramid in an organization chart. This hierarchy exhibits a high degree of specialization of function in various departments and a proliferation of formal policies, written regulations, job descriptions, and other prescriptions.

In bureaucratic organization employees are hired to fill the prestated requirements of particular jobs; personnel are presumed to be interchangeable within job categories; and thus men are hired to meet the requirements of the job, rather than tailoring the job to correspond to unique individual capabilities. Furthermore, the normal pattern of career progression is upward through the heirarchy into managerial roles. Finally, and perhaps most important, the bureaucratic form of organization ordinarily assumes that quality control over work will be exercised by higher level management officials within the organization, rather than in terms of internalized professional standards.

Thus contrasts have emerged between the conditions of work that scientists have learned to expect by virtue of their professionalized training and the conditions of work that they find in the more highly bureaucratized environments of government and industry. Conflict in such situations may be fairly predictable, but it is not irreconcilable. "Adaptation mechanisms" have enabled scientists to live in these strange organizational environments and, in turn, have enabled different kinds of organizations to utilize scientific products in ways that support organizational objectives and scientific interests at the same time.

One class of mechanisms, called "professional adaptation mechanisms," permits scientists to pursue objectives in accord with professional values and modes of behavior in organizational contexts like government and industry. Scientists have been taught to value basic research. They expect to do their work with a bare minimum of direction from nonscientists; they tend to be especially jealous of what they consider to be scientific prerogatives, including a high degree of freedom in choosing the research problems they engage in, as well as a high degree of freedom in the actual day-to-day conduct of research.

How do they they go about trying to achieve freedom in organizations that aim chiefly for defense, public service, or products, but not for the advancement of science? Scientists most frequently achieve freedom by "status-advancement," that is, moving into supervisory positions where scientists can control research programs. This shift creates a dilemma. Although scientists obtain more control over their work, they must concentrate more on administration and less on conducting research. First-line supervisors or research group heads and lower-level laboratory managers can conduct more research, only at the expense of administrative responsibilities. Moreover, they become the "man in the middle," simultaneously representing managerial requirements to scientist-employees and employee interests to management. Many do not like such duties.

A less common way for scientists to achieve freedom is by means of "research entrepreneurship": selling research ideas to sponsors—individuals or groups who provide funding—either inside or outside the employing organization. The relation between a scientist and his sponsor becomes very much like the institutionalized relation between any professional person and a client. The significance of the research control is profound because it establishes a relationship that cuts across the normal lines of hierarchical authority in a bureaucratic organization. The scientist, as principal investigator, becomes responsible for the quality of his work to a project monitor of the sponsoring agency rather than to his immediate administrative supervisor.

Another mechanism available is "bootlegging": engaging in non-formally specified research activities within the larger formally structured projects. Bootlegging sometimes is done with a sense of guilt, but more frequently does not involve evasion of job responsibilities. Usually work statements are general enough to give scientists this flexibility. For their part, organizations have developed a variety of ways to integrate the activities of scientists into their missions. These are called "bureaucratic adaptation mechanisms." The first of these is recruitment. Organizations try to hire employees best suited to organizational desires and fire those who are not as suited. Problems of recruitment are great. Employers often believe that a doctor's degree is preferred among scientists. Conversely, in applied research, holders of master's degrees may make more appropriate workers.

Furthermore, scientists without doctorates tend to identify personally with an employer and plan for a long-term career with him.

After serving for a while, most employees begin to develop a better appreciation for the objectives and requirements of their employer and to personalize his goals as their own. This is known as the process of organizational socialization. In this process, scientists, like other employees, tend to strongly identify themselves with the organization. After 1 to 5 years, however, comes a period of disillusionment as they find that few organizations look as good to insiders as to those on the outside. Then, as the employee learns the "ropes," he learns how to live within the organization and again becomes attached to it.

This process raises a number of important questions. Does management really want scientists to remain long enough to become committed? Are individuals who retain strong professional scientific interests more likely to leave, gradually increasing the proportion of older, less interested and less capable scientists?

Another question relates to incentives—inducements and rewards to keep the scientist tied to the organization. Salaries are as important to scientists as to other men. They will move to organizations that pay their colleagues more money for similar work. Most enlightened managements try to provide high salaries for scientists and technical people. Some managers make the mistake of assuming that higher than average salaries will necessarily attract better than average scientists. On the contrary, several studies have indicated that once scientists have what they consider an adequate salary and fringe benefits, nonmonetary incentives, especially freedom to select their projects and laboratory facilities, count more.

Thus the professional orientation of scientists—that is, the values that are ingrained into them in their long period of professional preparation and education—emphasizes the importance of autonomy and freedom in the conduct of their work. They carry this orientation with them when they become employed in industry, government, universities, or elsewhere. The professional values and disciplines that they have acquired are almost as strongly ingrained as the codes and disciplines of the professional soldier. If organizations are to utilize these professional scientists in an effective way, they must provide for the fullest expression of professional attitudes and activities, rather than resist them. Bureaucracy and professionalism must accommodate to each other.

The research manager must balance the controls required to attain his objectives within his limited resources against the freedom which roving intellects need for untrammeled creativity. He also must take account of the scientist's questioning attitude toward institutional policies and procedures, of his insistence on freedom, of his distaste for being managed or taking on administrative responsibility, and of his dislike for the diversionary influences upon him.

These characteristics not only plague the manager with thorny problems but also offer him potential assets. These personality traits of the scientists aid the manager in several ways. The researcher is self-motivated toward his chosen goals. Furthermore, he is concerned with creative contributions toward broad programs just as much as with individual projects. The great challenge in the manager-scientist relationship entails more than making the researchers more management-minded and management more tolerant of scientists; it also involves the demanding task of integrating these two attitudinal requirements into a higher administrative art.

These characteristics make the research laboratory a special institutional creature. Clear definitions of programs and plans become particularly troublesome. How can one estimate the time and cost of formulating a new complex of ideas? Although he is constantly striving to improve his planning, the manager must live with inexactitude. In this situation the imaginative manager may find need to develop institutional strength through the creative use of competition and conflict.

In addition to the professional attitudes of the scientist, research management problems are also affected by other factors such as the recent explosive growth in the size and complexity of R&D programs; the accelerating frequency of program changes with its interruptions to continuity; the speeding up of technogical obsolescence; the shortages of trained subprofessional manpower leading to a growing dependence upon higher than necessary levels of competence; the cost of R&D rising at a rate faster than general inflation; and the critical interest and the overstimulated expectations of legislators and the general public. While these and other factors bear on research administration in unique ways, it should not be forgotten that in many respects research management remains similar to other forms of institutional management, especially since less than a third of laboratory staffs usually consist of the professional scientists.

Leadership

Although often called "director," the chief technical head in a successful research activity tends to operate as a senior among colleagues rather than by managerial directives. As with other executives, he must deal with both external relations and internal management. The research executive must make decisions on program areas and priorities; attract staff interested and competent in these areas; provide resources and environmental conditions conducive to innovation and cooperation; assess progress; and do all these things by providing foresight. Within the research activity, he is more likely to act as catalyst and inspirer—as the orchestrator of communication and collaboration, and at times, final decisionmaker. The research manager probably uses committees and other forms of group participation extensively at all levels. As a rule, he delegates routine business and facilities-maintenance functions to an assistant for administration.

The head of a research establishment is, with few exceptions in the United States, a scientist or engineer himself. Usually he has come up through the ranks. In any case, he often has obtained stature in some aspect of the technical work which he oversees. Otherwise, he faces severe handicaps in attracting significant program responsibility and a competent staff; in interpreting requirements within higher councils, and in protecting his organization against misuse. Moreover, he can provide a good working environment and effective internal leadership most readily if he knows the language and ways of scientists and engineers.

Do the most eminent scientists make the best executives? Certainly, their stature provides a great asset, at least in meeting the technical staff's psychological needs. But many scientists dislike administration. Moreover, those who achieve scientific renown frequently possess the vices as well as the virtues of genius.[3] For a scientist-executive, assets other than scientific eminence count more in the long run. Among such attributes are the ability to work with laymen; the will and skill

---

[3] E. S. Hiscocks, *Laboratory Administration* (London: Macmillan & Co. Ltd., 1956), p. 47.

for explaining the results of his work; a generally well-developed verbal ability; an interest in scientific and nonscientific fields other than his own; an ability to "size up" people; a respect for the ideas of fellow professionals; an appreciation for administrative considerations; skill in the effective use of the talents of subprofessionals; adroitness in committee work; possession of the respect of his colleagues; a willingness to forego the depth of analysis he practiced as a professional scientist; and a temperament to settle for imperfect solutions when necessary.

In a negative vein, the research executive should not be overly sensitive in his relations with people; unduly upset by delays or changes; impatient about having others review or coordinate his work; or be interested in managerial work only to further his real professional needs which lie in other directions.[4]

The research executive may develop by his gradual growth from full-time researcher through manager-scientist to full-time executive. Higher management selects him for this development by evaluating him in his committee work, by giving him special staff assignments and by allowing him to step into executive roles during emergencies. By such means, the scientist also tests his interest in executive work. It has often happened that the moderately successful scientist proves to be a better candidate for executive responsibility than the most highly productive and creative scientist.

Scientists with managerial potential often manifest an initial aversion toward accepting executive assignments. This can be overcome. Such aversions are part of the status stereotypes among scientists and, consequently, do not necessarily represent true personal feelings. To overcome reluctance it should be made clear that a research executive with proper administrative help can often continue some of his personal research. In addition, the right individual may discover satisfaction by multiplying his creative contribution through his leadership of others.

Although conscious planning for executive replacements is basic in developing science administrators, higher management sometimes fails to plan with sufficient vigor and care. Rather than feeling out a man's interests early in his career, coaching him in developing human relations skills, and giving him insights into the broader aspects of research

---

4 Milton M. Mandel, "Research Management: Some Clues for Selection," *Personnel* (January-February, 1961), p. 74.

administration, there is a tendency to avoid the challenge. Good management should maintain an inventory of anticipated vacancies and regularly assess potential talent for executive replacements. Furthermore, it would be desirable to have tailor-made rather than patterned approaches to indoctrinating supervisors and broadening executives as they move up the line.

Not all that an executive needs can be "built in"; some of what he needs seemingly must be "born in." Yet, it is possible to inculcate or strengthen certain abilities. The executive can be educated to see problems in broad perspective and to make decisions which serve long-run in addition to short-run goals. He can develop the capacity to delegate functions in such a way that authority is commensurate with responsibility. At all times he should maintain an open-minded receptivity to suggestions and comments from peers, subordinates, and superiors. He can increase his willingness to accept risks and criticism by thinking independently and taking firm stands when necessary. Great importance should be given to aiding the executive in developing a knack for discovering and utilizing previously undetected relationships within his environment as well as in improving his competence in simultaneously carrying on, integrating, and coordinating varied interests and activities. One of the best means of executive development is through systematic discussion of problems and circumstances among colleagues.[5]

Growing numbers of top-level executives in the federal government have been scientists or engineers. In many cases their responsibilities extend beyond direct involvement in research activities. As technological programs in government grew, we witnessed scientists and engineers assuming even more prominent roles among the nation's executives.

Planning and Programming

To recognize what is possible and worthwhile becomes the central problem of a director of research. If he restricts a staff scientist's choice of work, he necessarily interferes with the scientist's freedom in exploring the unknown—the very essence of scientific creativity. A

---

[5] Robert N. McMurray, "Man-Hunt for Executives," *Harvard Business Review* (January-February, 1954), p. 13.

scientist bent on pursuing the "gleam in his eye" usually lacks enthusiasm for substitute projects. It is, therefore, a prerequisite that the director select researchers whose interests harmonize with those who design the program. He should then consider in his plans the knowledge and inquisitiveness that his scientists bring to their work plus their commitment to mutual interchange with their colleagues. Whatever else the effective manager does in his planning, he must seek to release and utilize the knowledge of his professionals in their specialized fields.

Able management planning first satisfies the need for defining the mission of the research establishment including its main purposes and roles. It then seeks to lay out a scheme for making more effective use of researchers with managerial talents and for improving the competence of management staffs by providing opportunities for them to become better versed in science. Organizational planning also takes into account the tendency within research activities to utilize technical personnel in functions quite distinct from their training, ranging from "parts chasing" to top management. This tendency at times bleeds off a considerable amount of scientific talent which could be used more profitably in other ways.

Among his other concerns, the research manager should provide for the reorientation of the scientist. One important benefit would be the encouragement of the scientist to recognize the value of the professional manager. The latter can relieve the scientist of many of his non-research chores thereby significantly increasing the research payoff from available resources. By and large, the scientist should become a research manager only when he honestly feels that he can thus contribute to improving decisions with regard to the technical work.

The parent agency's reasons for undertaking research no doubt govern research planning at the activity level. However, in order to prove a meaningful guide, the mission statement should be formulated in reasonable detail. The mission statement should help those outside a research facility to relate their work to the facility's purposes and to use its services. Even though it may be stated in broad terms to allow latitude, the mission statement requires review periodically to insure timeliness and usefulness.

The research plan contains the main lines of attack to achieve the goals implicit in the mission. The plan involves projecting future needs; judging the more promising opportunities; assessing present and estimating future resources; and identifying priorities. In progressing from the more general to the specific, the planning includes long-range, intermediate, and immediate objectives. Shifting to more detailed planning usually does not run smoothly in this sequence. One cause for trouble in research planning, as well as in other activities, is preoccupation with the immediate to the neglect of the long range. This may lead to excellent planning and performance of something which the research activity should not be doing.

The powerful influence of short-range views is understandable if we consider the inherent difficulty of planning research in any detail. As one moves from fundamental research to development and engineering, planning becomes more feasible, although in the latter types of activity planning still can be somewhat imprecise.

Plans are generally given specific and concrete form by means of a program comprising individual projects, tasks, and subtasks. While some projects may originate with top management or outside the research organization itself, in the more basic areas of an activity's research mission the researchers themselves most often make the proposals. Even if the suggestion originates elsewhere, the scientists themselves should conduct the preliminary search of the literature and do the exploratory work to relate need to opportunity. The documentation of each project usually covers objectives and technical background, the rationale for the undertaking, specific tasks keyed to time intervals (with the greater detail for the near-term time period), and the immediate and future need for material and personnel resources.

Project proposals normally receive supervisory review as well as review by top management committees or councils which examine the entire program, including the finances. Once approved, the projects and project summaries generally serve as one basis for budget justification and fund allocation. Project summaries also serve as valuable communication and reference tools. The research activity should decide whether to distribute this information selectively in-house or among its entire staff.

The director's judgment constitutes the single most crucial factor in determining the success of research programming. [6] No programming

---

[6] Mees and Leermakers, p. 212.

system can replace able leadership and clear decisions. The sensitive leader recognizes that the researcher tends to dislike planning, that is, committing his intentions to paper. The director balances the possible restraint which structuring can impose on freewheeling creativity against the recognition that the average researcher produces more effectively when he has clear goals and accountability.

In judging a project, the executive considers the need for prior knowledge in the area of the proposal, its urgency, the probability of fulfilling the stated need, the available resources, and the consequences of success or failure. He also looks for clearcut guidance in statements of general policy and arranges for preliminary investigation before making commitments. It is his responsibility to coordinate each project with related projects. He should also explain the reasons for turning down proposals in a way that encourages further proposals. Another aspect of his concern is to review significant changes in previously approved projects.

Planning the termination of a project often proves troublesome. It is a recognized fact that research frequently continues after it should be ended. Instead of outright termination, managers sometimes temporarily shelve a project until the need for resumption becomes evident.

Much of the research in the United States today is carried on by private contractors in behalf of sponsoring organizations, mainly the federal government. This relationship leads to some special problems. As one example, industrial research contractors sometimes prepare overly elaborate and costly presentations of proposals. They often utilize some of their best talent for this task and thereby reduce the time these men of talent actually devote to research. Another problem is that the government, as the contracting agency, may not have sufficiently competent liaison personnel. The government contracting official, if he himself does little research, soon loses his capability for planning and evaluating research projects. He also may tend to surround the project with unduly hampering restrictions. These problems challenge the governmental manager of research programs in our system of "federalism by contract." [7]

---

[7] Don K. Price, *Government and Science* (New York: New York University Press, 1954), pp. 20, 65.

Organizing the Research Activity

The proper organization for a given research establishment depends upon such factors as the size of its programs, homogeneity or diversity of its activities, the number of its leaders with their preferences and styles, and, often, the wishes of the parent organization. An organization structure provides a framework that subdivides functions into meaningful parts, sets focal points of accountability and leadership, and identifies formal lines of communication. Yet, it presents but a skeleton of a complex social organism, and, as with an X-ray, sketches only an outline of what the structure really is. In a research group, the informal organization is likely to be more meaningful than that depicted in the formal chart. Formal organization structure is most effective when it facilitates rather than frustrates the group in carrying out its basic functions.

The organizer of a research activity faces perplexing dilemmas. He must retain the advantages of small research groups within a large organization without piling layer upon layer of supervisors in the pyramid. A partial answer, especially in the more basic types of research, has been to choose a wide, flat structure. Many project leaders report to each section chief, many section chiefs report to each division chief, and many division chiefs report to the technical director. Such a spread in span of attention appears preferable to imposing more than three levels of review upon the "bench scientist."

Subdividing activities poses other choices. Experience has evolved two general methods for structuring a research organization—by function and by project. In the functional method, the major breakdown may be according to major scientific discipline (physics, chemistry, biology), followed by further division either by general field of interest (radio propagation, temperature, polymers) or by stage of investigation (basic or applied research). The essential characteristic of the functional method of organization is grouping in terms of expertise.

The project type of organization, on the other hand, brings together into a single working group the contributors from different specialties. Besides being multidisciplined, the project group often is an ad hoc organization for solving certain assigned problems. In practice, however, some of these groups may prove to be long termed. Nearly all research organizations have found that combinations of functional and project organizational arrangements are necessary, especially in large undertakings.

Functional organization by discipline offers both advantages and disadvantages. On the plus side, it pleases the scientific professionals by emphasizing specialization and fostering the development and use of individual skills. By keeping a scientist close to his specialty, it permits clear-cut advancement ladders and performance standards. It also tends to bring together a compatible family of colleagues.

On the minus side managers must guard against the intellectual parochialism that specialists may develop if left alone. Steps are needed to prevent professional goals from overshadowing organizational goals, and colleague approval from outweighing management approval. There is some tendency in functional organizations toward stodginess and promotion of vested interests as well as some evidence of insecurity due to the obsolescence of one's research specialty. To these dangers can be added a possibility of organizational rigidity; a magnification of coordination problems; an increase in scheduling difficulties; a proliferation of service groups; a growth in difficulty in evaluating progress and in terminating projects; and a general tendency to work on ever more specialized problems of ever decreasing significance.

Organization by projects also offers advantages and disadvantages. Among its merits, it provides an interdisciplinary "mix" of brainpower focusing on a single problem. In so doing, the research team acts as a cohesive social unit in exploring a definite scientific question. Orientation toward goals is stressed. Team members become committed more easily to one another and learn to appreciate the other fellow's specialty. The diversity of experience of the project team members broadens each one's perspective and makes for versatility and flexibility. As a consequence, innovation may result more readily.

Furthermore, organization by project, emphasizing vertical structure, tends to stimulate enthusiasm and dedication. Because the members of the team can focus on a common goal, a cooperative spirit may produce unexpected accomplishments. The groups use subprofessional help more frequently and are less plagued by "lone wolf" isolation. The manager can assess as well as budget the entire project more effectively and, when terminating or redirecting a project, he upsets one rather than many groups.

On the debit side, a project-oriented group tends to become overly self-contained. This may limit the use of the available experts in other projects. There may also be evidence of intergroup jealousies; use of amateurs to handle technical matters instead of specialists with the needed skill; "empire building"; inadequate opportunities for communication among professional peers; and group unity achieved at the expense of alienating the rest of the research activity. Managers also may find that the diversity of the people and the temporary character of project units accentuate incompatibility among the members.

Other undesirable consequences may occur in project-centered groups. Stimulus toward creativity in one's own field may be reduced as the researcher inclines to becoming a jack-of-all-trades. Standards of evaluation of individual peformance become confusing and the organization lacks a clear line of advancement in the individual's specialty. This fact plus a lack of a permanent base causes some insecurity for the individual. The group also tends to depend heavily upon an effective project leader, a relatively scarce person. Furthermore, management must constantly adjust the size of the group as changes occur at various stages of the project. Experience tends to show also that duplication or underutilization of equipment is more likely to occur in this type of organization.

The project-type organization may pose other vexing problems. Insufficient communication among specialists in the same discipline may result in the rediscovery of things already known, with a consequent waste of resources. In addition, a group having a concern only for a specific project may tend to perpetuate itself when termination would prove more appropriate.

Communication of Technical Information[8]

The main product of research is scientific and technical information. Researchers make their own product known to others by outward communication while they utilize incoming communication concerning

---

[8] A far-ranging study of the problems of communication of scientific and technical information was undertaken by the Committee on Scientific and Technical Information of the U.S. Federal Council for Science and Technology. See *Recommendations for National Document Handling Systems in Science and Technology* (Washington, D.C.: Clearinghouse for Federal Scientific and Technical Information, U.S. Department of Commerce, November 1965). This document includes a two-volume background study as appendix A.

the results of work done elsewhere. The transfer of information among researchers is a most important aspect of the conduct of research.

Recognizing this fact, the alert manager takes certain steps. He first identifies the important areas of specialized knowledge of interest to his activity and assigns the appropriate specialists to keep up with these fields. These specialists often need to organize technical data centers which they should share with others in the activity. He encourages outside contacts, including attendance at professional meetings and taking sabbatical leaves. He also arranges to provide library aid through book and periodical collections, and, especially, staff help in processing and evaluating incoming material. Many research managers find that it pays to devote competent talent, such as junior scientists, to this service. The library should promote the dissemination of information as well as respond to requests. Even with these aids, the researcher often finds himself unable to cope with the information he needs to keep up to date. The conclusion becomes inescapable that no researcher can fully keep up with all the pertinent, original literature in his field. Consequently, there is disconcerting evidence that some research only results in rediscovering what is already available in the literature.

Research managers have promoted possible solutions to the scientific information explosion. A whole body of knowledge, if not a whole profession, is developing to handle the problem of information storage and retrieval of large quantities of technical data. Practical aids, some using computers, and consultants are now available. Managers must be wary of falling victim to pressure to adopt novel but overly complex systems or unnecessary gadgetry. Moreover, no matter how excellent the information finding and selection aids may be, many researchers still need to read more efficiently. To meet this need some research activities have found it worthwhile to conduct training in reading improvement.

Internal communication within the research activity also poses important challenges to management. Upward communication should function smoothly so that findings at the bench can influence policy at the top. To energize the upward flow, research directors should maintain a constant informal relationship with their working staffs and thus demonstrate their continuing interest. More formal aids to communications include the preparation of progress reports, circulation of records of projects, and program reviews. In addition, management should weigh the value of organizing staff meetings, technical symposia, and informal discussion groups. Furthermore, the physical arrangements of the research activity as well as the general climate can facilitate internal communication.

Except in certain special cases, a crucial function of a research establishment is to disseminate its research results to the broader scientific community. Many government laboratories consider this as one of their chief services. Contributing to the general technical literature is a prime objective of most scientific groups. Besides technical publications, other channels of dissemination include reports to sponsors, holding open houses, exhibits, films, television presentations, lectures, correspondence, reports to congressional committees, public addresses, and news releases to the technical and popular press.

As may be consistent with its functions, a research establishment often, in addition to engaging in outside publication, publishes its own books, journals, or monographs. For example, the National Bureau of Standards (NBS) publishes several periodicals such as a journal of research, containing contributions to the fields of physics, chemistry, mathematics, engineering, and instrumentation. NBS also distributes nonperiodical literature in the form of handbooks and technical notes.

The amount and quality of formal publications determine to some degree the prestige of the research establishment itself. Considerable attention, therefore, must be paid to its processes of review and approval. Reader panels and editorial boards are common although they may prove cumbersome; the researcher sometimes becomes frustrated by long delays in seeing his "brainchild" in print. To overcome some researchers' difficulties in providing well-written manuscripts, manuals on good writing as well as individual and group coaching in writing skills may be helpful. In addition, more and more research executives are furnishing expert technical writers to transform the scientists' output into more readable form.

A very vital problem involves communicating appropriate information to nonscientist executives, legislators, and the general public. The effective research director devotes some of his time and energy to increasing public understanding of the frontiers of science. By fulfilling this public responsibility he aids his own cause because a better informed public more readily supports scientific endeavors.

Development: The Application Phase

A development effort is the medium by which knowledge and techniques are transformed into useful devices or systems. The degree to which scientific methodology contributes to development efforts varies from project to project. Some projects may be called "applied research" and many of the management characteristics of such projects are similar to those of research itself. Other development projects heavily involve manufacturing aspects. For these, some of the management characteristics may be like those of the production process. Overall, however, the development phase does have a number of unique characteristics, differentiating it from either research or manufacturing.

First, let us cite the major characteristics that development management shares with research management. Development also places great emphasis on innovation, needs an effective two-way flow of information, and requires leaders well versed both in managerial skills and technical knowledge. There is a similar need for designing an organizational environment in which working scientists and engineers have an opportunity to exercise creative imagination.

There are also important differences. The manager of a development program usually must produce a system or part of a system meeting a stated need (in terms of reliability and performance) within a certain time period and at a certain cost. Consequently, he faces an engineering more than a research task. Rather than attempting to discover new basic physical laws affecting his end product, he generally uses those already known in order to create a useful system. His basic purpose is to accomplish an agreed-to end objective. While he may take some risk to gain an original breakthrough, he usually proceeds systematically in building the technology step by step. With a specific goal in mind, he cannot allow his engineers untrammeled freedom in their activities, but rather must integrate a number of subprojects which together result in the desired end-product.

The application of knowledge and theory to the solution of stated problems begins with fashioning pertinent techniques, conducting feasibility demonstrations, and creating experimental components. Next comes the development of experimental assemblies and, later, full-blown prototypes. Throughout this progression, managers must arrange for the resources suitable to each step in timely fashion.

Faced with specific time, cost, and performance requirements, managers of development programs have increasingly come to use sophisticated management techniques. These techniques are essential in large-scale development undertakings. Perhaps the most celebrated of these management techniques is PERT (program evaluation and review technique). It is designed to permit detailed planning, analysis, and evaluation of the cost and time required to complete the steps in a major development. It provides a basis for early identification of potential schedule slippages or cost overruns so that corrective actions can be taken. In some applications, PERT employs computer processing. First used in developing the Navy's Polaris system, it has been adapted for use in a wide variety of large-scale technological programs. Depending on program needs, PERT and other related management tools may be used to assist the manager in reaching his objectives, especially in controlling time and cost of development programs.

Summary

The increased emphasis upon systematic application of knowledge and techniques to further a new understanding of nature and useful products has created a demand for improved R&D management philosophies and practices. In many instances, management of R&D requires unique approaches; in others more commonly used concepts remain appropriate. Perhaps the greatest challenge facing the R&D manager is determining when and where special interpersonal and institutional relations will prove most effective, and when and where they will not. In addition, the R&D manager has the responsibility of improving the special managerial instruments as well as general managerial devices that apply to R&D.

4 Major Research Frontiers

Ralph Sanders

Forecasting the state of the art in scientific and technological enterprise is fraught with uncertainties. It was George Eliot who felt that among all forms of mistake, prophecy is the most gratuitous.

While it may be hazardous to attempt to prophesy gains in specific fields, we can confidently predict that overall scientific and technological progress will occur. Examining the goals and interests of present-day research yields fairly reliable clues as to where it is most likely to emerge. Breakthroughs remain unpredictable, but more often than not they happen in fields that receive the greatest support in men and money (this relationship is especially true in technology). History has shown, moreover, that science does not advance by a series of sudden discoveries. Usually, there are long periods filled with uneventful, but nevertheless very important, drudgery. The gradual accumulation of new data leads to new ideas requiring new instrumentation using new techniques. Yet, the actual unfolding of this process is not always immediately apparent.

This chapter covers chiefly those frontiers which U.S. researchers are emphasizing. Certainly other nations direct some of their research programs along similar lines. In many cases, foreign countries adopt different approaches or look for different results. For instance, the Soviet Union has developed a termination system for manned space flight on land while the United States returns its spacecraft over water.

Mindful of these qualifications and reservations, this chapter attempts to identify some scientific and technological endeavors that seem likely to produce significant advances in the foreseeable future. The selection is intended to be neither exhaustive nor all inclusive. Nor is there a universally accepted judgment as to the amount of discussion that each of these frontiers deserves; other observers could very well have preferred to have enlarged or shortened the coverage devoted to any one. At best, this selection attempts to highlight the more promising lines of research and engineering in the United States; no effort is made to rank their importance.

Space

Among recent achievements in science and technology none excited popular imagination more than the exploration of space. The National Aeronautics and Space Administration (NASA) launched a family of multipurpose satellites and deep space probes. Each carried a large number of instruments designed to make coordinated observations in a number of research problems, especially on solar, geophysical, and astronomical matters. Orbiting solar observatories, for example, have been equipped with instruments for measuring solar X-rays that scientists cannot adequately study from the ground because of atmospheric distortions.[1]

Over the years the United States has conducted programs using earth satellites as communication relay stations and for weather observation posts. Some of these have made the transition from experimental to operational status (and in the case of communication satellites to commercial use). Unmanned explorer satellites have carried on scientific studies of regions surrounding the earth, the moon, and the sun. A biosatellite, for instance, carrying equipment to conduct complex biological science experiments has been orbited.

NASA has conducted a lunar exploration program. The lunar program investigated lunar materials, largely associated with the Apollo man-on-the-moon program. NASA scientists also seek to explore other planets of our solar system in order to learn more about the beginnings and development of the earth and related questions of the origin and evolution of life and man's terrestrial environment. The planetary programs employ ground-based astronomy facilities and automated spacecraft to fly by, orbit, or land on other planets. U.S. spacecraft have orbited Mars and flown by Venus and are planned to land on Mars. One craft flew by Mercury. A number of considerations has led NASA to forego the proposed "Grand Tour" mission to all the outer planets—Jupiter, Saturn, Uranus, Neptune, and Pluto—and instead to fashion a more modest program. NASA is

---

[1] U.S. National Aeronautics and Space Administration, *21st Semiannual Report to Congress: January 1 to June 30, 1969* (Washington, D.C.: NASA, 1969) and U.S. Congress. Senate. Committee on Aeronautical and Space Sciences. *NASA Authorization for Fiscal Year 1973*, 92d Cong., 2d sess. (Washington, D.C.: U.S. Government Printing Office, 1972).

redirecting its efforts to focus in the late 1970s on Jupiter and perhaps Saturn with less expensive spacecraft.

Although unmanned vehicles can effectively carry out many of these exploratory tasks, various types of missions call for manned spaceflight. Man has certain capabilities which machines may never match. He is unique in his ability to make on-the-spot judgments, discriminate between unanticipated alternatives, and adapt to rapidly changing conditions. Recognizing the value of these advantages and conscious of the major scientific and technological challenges from abroad, in 1961 President John F. Kennedy committed this country to placing a man on the moon within the ensuing decade.

In 1969, astronaut Neil Armstrong took the first exciting step on the moon, signaling one of the most impressive scientific and technological accomplishments in history. We still cannot comprehend the potential long-term impact of the fact that, for the first time ever, man trod upon a terrestrial body other than his own Earth. By the end of 1972, NASA's Apollo astronauts had made five successful trips to the moon, bringing the Apollo program to a close.

While completing the moon landing program, the United States begin to shift emphasis from high adventure projects to the more prosaic task of making spaceflight economic and contributing knowledge that might be usefully applied. Accordingly, NASA instituted the Space Flight Operations Program to stimulate direct practical benefits to mankind and reduce the cost of getting payloads from the ground to low earth orbits. This program's chief efforts are the Skylab and space shuttle projects.

Skylab, the first U.S. experimental space station, in orbit in 1972 and 1973, was the largest manned vehicle ever built. It provided living quarters and a laboratory for scientist/astronaut crews who carried out extensive experiments in biomedical sciences, earth resources, astronomy, space physics, engineering, and manufacturing technology. The space station's three 3-man crews remained in flight for extended periods, in one case up to 4 months. Skylab succeeded in gaining detailed understanding of the capability of human beings to function effectively in space for extended periods of time.

Perhaps the key to future United States space applications and explorations is the space shuttle. An entirely new way of getting to and from space, the space shuttle will be an airplanelike vehicle that can be used over and over to take satellites to orbit and bring them back for repair and reuse. It can perform manned missions in orbit, but its principal use will be to conduct missions with unmanned satellites more efficiently and economically than we now do with launch vehicles that are destroyed on every launch.

Each space shuttle consists of an orbiter and a booster. The orbiter looks like an airplane and is about the size of one of the smaller jet airliners. It will have a large cargo bay, 14 to 15 feet in diameter and up to 600 feet in length with a carrying capacity of some 65,000 pounds. It will have a two-man crew and carry 2 to 12 scientists and technicians. Its personnel compartment will be pressurized so that the crew can travel in shirtsleeve comfort without spacesuits. The booster will be an unmanned, ballistic-type vehicle, using either pressure-fed or solid rocket engines. The booster and orbiter stages will be joined for launch, with the orbiter in a "piggyback" position. At a predetermined altitude the orbiter will separate and fly in orbit like a spaceship. When it finishes its mission, the orbiter will land like an airplane and be made ready for the next flight.

NASA, the Department of Defense, and other agencies will use the space shuttle to carry out unmanned, man-tended, and manned missions (nonastronauts can participate in manned missions). These missions are designed for:

Scientific exploration—using instruments, observations, and telescopes to study the earth, planets, the sun, and the universe.

Civilian applications—missions involving weather activities, communications, navigation, monitoring the environment, identifying earth resources, and promoting new ideas for the future.

Military applications—missions that will take advantage of the shuttle's economical, routine, and quick reaction capability.

The economic savings should be great. The cost per mission ($7.7 million) is less than all existing launch vehicles except for the small Scout and Delta, even though the shuttle's 65,000-pound capacity is far greater than any of them (except for the giant Saturn V). The cost per pound in orbit when fully loaded is about $120 compared to $900 to $5,600 per pound for conventional launch vehicles.

The versatility and low operational cost of the space shuttle should enable it to perform the following specific functions effectively: (1)

placing satellites in orbit and returning them; (2) repairing and servicing satellites; (3) delivering propulsion units to low earth orbit for further launch of vehicles into high orbit or to escape trajectory; (4) short duration science and application missions; (5) serving as a space laboratory; (6) space rescue, and (7) providing support to space stations (called Spacelab) that will be placed in orbit in the future. The space shuttle is planned to be operational by the end of the 1970s.

Another illustration of this country's commitment to use space for practical purposes is NASA's Earth Resources Survey. Earth Resources Technology Satellites are launched into medium altitude, near-polar orbits to investigate earth resources relating to agriculture, forestry, geology, hydrology, cartography, geography, oceanography, and environmental quality. These satellites, using such equipment as high resolution television cameras, a multispectral point scanner, and a wideband video tape recorder, detect terrestrial and ocean characteristics that might indicate plant diseases, tides, potentially useful ores and minerals, and similar information that could prove economically and socially valuable.

Oceanography

U.S. attention to outer space currently is being accompanied by a growing interest in "inner space," that is, the oceans and seas of the earth. A study completed in 1964 by the National Academy of Sciences estimated that for an annual investment of less than $200 million, the United States could reap thirty times as much in economic benefit. While it is difficult to attempt to place a price tag on the fruits of research, it appears likely that eventually significant economic gains can be made by pursuing oceanographic research and development. The implications of this research for military, but especially naval, operations can be far reaching and result in more advanced weapon technology and tactics.

With all the scientific and technological progress during the past few centuries, it is surprising that men know relatively so little about the oceans and seas which cover more than three fifths of the earth's surface. Close to home the continental shelf adjacent to the United States comprises an area of about 850,000 square miles, about one fourth of the country's total land area. Almost 95 percent of the shelf lies at depths of 600 feet or less. Yet, we know very little about the potentially huge organic and inorganic resources that the shelf may contain. Commercial fisheries and oil companies, the industries most interested in the economic assets of the sea, have barely begun to probe its treasures. We are almost totally ignorant of the character and composition of the terrain beneath vast stretches of ocean, especially those distant from our shores.

Our theoretical concept of ocean dynamics is crude and in many cases uncertain. So far, we have only approximations of how the oceans behave and these fall short of the reality. Scientists, through experimentation, pursue a goal of predicting oceanic phenomena and of refining the major theoretical tools developed in the past 20 years. This theoretical base explains the major features of the ocean and its currents, but has not yet been thoroughly checked experimentally in the deep open ocean.

This nation has been accelerating its investigations into the nature of the oceans. In the spring of 1965 the Navy initiated an expanded 3-year oceanographic and geophysical survey of the world's oceans. Other government agencies undertook related programs. Increasingly we are placing more emphasis on the actual exploitation of the seas. Ocean engineering, composed of many kindred disciplines, is emerging as a significant industry.

The Department of Defense, especially the Navy, has a great interest in oceanographic technologies. Primarily, it has been trying to improve anti-submarine warfare devices and techniques as well as the current generation of ships, submarines and related equipment. Development of better deep-submergence vehicles, enabling rescue of personnel near present submarine collapse depths, is also urgent. The loss of the nuclear-powered submarine *Thresher* in the North Atlantic in April 1963 and the Navy's difficulty in locating the wreckage and inability to raise it, stimulated intense studies to develop effective deep submersible vehicles and operating procedures. The difficulty in recovering the H-bomb lost off the coast of Spain in January 1966 reinforced the belief that additional research and development in deep-diving vehicles are needed. Such vessels and proposed undersea platforms should enable men to explore about 90 percent of the ocean floor.

The tasks of conducting naval operations at greater depths, of exploring the oceans, and the mining of valuable products from the seas and the floors beneath them require a fashioning of many interrelated technologies, including devices for location, research and salvage; new sensors; new hull materials; structures and undersea construction techniques; salvage huts; improved rubber suits; swimmer propulsion units; improved deep-sea cutting and welding techniques; improved navigation, control and communications systems; new lightweight nuclear and fuel cell power plants; life support systems; and improved manipulators. Ocean engineers will require new and improved technologies for both submersible and surface vehicles.

Many complicated ocean floor operations have become feasible and are being investigated. However, if the new diving capability is to be fully exploited, new personnel equipment must be available to improve man's work capacity. Examples of such new equipment are more efficiently heated diving suits, specialized hand tools, bottom crawlers, mechanical arms or manipulators, more precise metering devices for breathing gases, and materials for bottom stabilization to decrease the amount of silt suspension in work areas.

The economic benefits of oceanographic research and development lie in such diverse fields as undersea mining, ocean shipping, fishing, sewage disposal, and recreation. The case of ocean mining illustrates both the promise and the problems of ocean engineering projects. Although in certain cases the possibilities of ocean mining do appear promising, many observers consider vein or lode deposits situated under the ocean floor as economically unattractive. Current research suggests that the extreme difficulty in locating and evaluating such deposits combined with intrinsically expensive manned undersea operations preclude early development of an undersea vein or lode mining technique. More feasible is the task of locating deposits on the ocean floor (mostly manganese nodules), and developing the technology for recovering and processing their ores. Progress has been made in fashioning the hardware for sampling deposits on the sea floor. Improved versions of the dredge offer dependable sampling mechanisms as does coring machinery. Researchers believe that large areas of the Pacific Ocean meet the environmental requirements for the formation of exploitable manganese nodules. Closer to home, the Blake Plateau, off the costs of Georgia and Florida, contains some low grade nodules that, unfortunately, are believed to be of marginal economic value.

Energy Resources

When in 1973 the Arab oil-producing nations imposed an embargo against the United States (and certain Western European countries) in the aftermath of the October war between the Arabs and Israel, this country came to an abrupt realization that it faced a major shortage of energy. Over the years, the United States increasingly came to rely on oil to fuel its economic growth because this fossil fuel proved relatively cheap, was more convenient to use, and more easily met the country's increasingly stringent antipollution standards.

By 1973 with 6 percent of the world's population, the United States consumed 35 percent of the world's annual supply of energy. It had far outstripped it own environmentally acceptable resources. In 1973 the United States consumed about 18 million barrels of petroleum daily in one form or another and expected this consumption to rise to 20 million by 1974. Its production could no longer meet demand. The gap could only be made up by imports from abroad, reductions in demand or increased domestic production. President Nixon called for the United States to achieve self-sufficiency in energy, an effort he labeled "Project Independence." If successful, Project Independence would by the early 1980s take this country to a point where it no longer would be dependent to any significant extent on potentially insecure foreign supplies of energy. Whether we meet this goal, only time will tell.

To meet its energy needs, this country first has to conserve energy by eliminating nonessential use and by improving the efficiency of energy utilization. In the first instance, increased energy efficiency can be achieved both by changing consumption habits and by applying existing technology. Industries, for example, have undertaken such measures as tuning-up plant equipment and insisting on more diligent management practices in plant operation; they are being more careful in their use of lighting and air conditioning. By using effective insulation and combustion control equipment, a great deal of energy could be saved in the operation of heat treating furnaces. The use of extant computer controls in operating large thermal processing plants likewise offers an attractive way to save fuel and reduce costs.

The nation needs R&D for realizing longer-range conservation possibilities. Researchers are developing a heat pipe vacuum furnace with energy savings in mind. Such a design that eliminates electric power generation associated with current furnaces, should reduce fuel consumption significantly. Housing design and construction techniques are other areas that offer considerable room for progress. Heavier construction and resultant increase in the thermal mass of a building reduce the peak load requirements for thermal conditionings.

Enormous savings in fuel can also be achieved by increasing the efficiency of the automobile engine. Presently, many engines operate at only 15 percent efficiency. Researchers are now studying the basic fuel combustion and conversion process. By gaining a better understanding of this process, researchers hope to boost engine efficiency to as high as 50 per cent in the next 15 years. Specifically, they want to expand their knowledge of the microscopic properties of matter and large-scale behavior of the combustion system. They are looking at staged combustion and diesel combustion characteristics, particle and droplet dynamics, fuel injection and control systems. R&D in this and other areas could prove very helpful in enabling this country to make better use of the available supplies of energy.

As we strive to conserve energy, we also are employing our R&D capabilities to expand our energy supplies. We hope to do so by developing technologies that would help us exploit more effectively resources such as domestic oil, coal, shale, geothermal sites, solar energy, and nuclear power. While these lines of inquiry do not represent the totality of U.S. energy R&D efforts, they do constitute the nation's major ventures.

Our efforts to find more domestic oil could be greatly aided by improving our oil exploitation technologies. For example, over the past 50 years there has been no major innovation in drilling. By stressing the development of more effective drilling technologies, we hopefully will produce major improvements in our ability to get to less accessible domestic oil formations and to facilitate the flow of such oil into our energy inventories.

The United States has huge reserves of high quality coal, conservatively estimated at 150 billion tons economically recoverable. Yet, this coal suffers from two major drawbacks: (1) it is in solid form and hence more difficult to transport and handle than gas or liquid fuels, and (2) a great deal of our coal has a high sulfur content, a dangerous pollutant that American society is intent on banishing from its air. As a result, in the past, energy producers preferred to use low sulfur oil (often from abroad) or go nuclear rather than use their resources to convert coal to a more convenient form and clean it up. In addition, moving coal R&D from the laboratory to commercial application proved difficult and expensive, lessening incentives for vigorous efforts. It was not until the price of competing fuels, especially oil, began to soar in 1973 that the federal government and energy producers began to think seriously of moving toward large-scale use and conversion of coal.

To overcome the sulfur problem we are now developing improved gas scrubbers, especially within smokestacks and exhausts, for sulfur removal. However, the technology still suffers from major operational and maintenance drawbacks. More R&D efforts are needed in this area if we are to meet this environmental challenge.

We also are directing our R&D efforts to derive synthetic fuels from coal to replace and augment our dwindling supplies of natural gas and oil. Researchers are looking to develop three major products—low BTU "power gas," high BTU quality pipeline gas, and liquid fuels including precombustion removal of pollutants. Actually, converting coal to fuel gas is a 19th-century technology, driven out of the American market by natural gas. The task is to come up with improved conversion systems that meet the needs of modern societies.

One relatively quick way to make more natural pipeline gas available to homes and businesses would be to replace the natural gas used by gas-fired electricity stations with power gas derived from low-sulfur Western coal. Several designs for producing power gas are being worked on. One of the more exciting is the fast fluidized bed for gasifying coal with air at high pressure to produce low BTU fuel gas to fire gas turbines. While power gas is not suitable for homes or businesses, as indicated, it is fine for power generation. Most observers believe that this innovation can be accomplished with existing technology, as quickly as plants can be built.

Developers also are fashioning a gasifier that operates at high temperatures to create a maximum amount of methane, a high quality gas fit for pipeline use. After being cleaned, this gas undergoes catalytic methanation wherein carbon and hydrogen atoms are used to produce additional methane. Another complex technique is being utilized to raise low BTU gas to pipeline quality through methanation and then introducing pure hydrogen into either the gasification or methanation stages for further enrichment.

During World War II, Germany made gasoline from coal despite its very high cost. At present developers are looking at fashioning coal-to-oil plants that combine in a new way nearly all the basic techniques used in coal conversion. The process is based on fluidization. Reactors of this system produce a mixture of hydrocarbons that are cooled and separated into two streams. The first produces medium BTU gas while the second results in oil factions that are hydrogenated into crude oil.

All R&D efforts to design commercially feasible conversion processes are directed toward three major objectives. First, the power gas, pipeline gas, or oil products must be economically competitive with other energy sources. Second, coal conversion systems must produce cleaner fuels acceptable to society. Developers are fully aware that it is easier and more economical to remove pollutants at the source of conversion than later from smokestacks or from the atmosphere. At the same time, equipment has to be designed and fabricated that withstands corrosion or the tendency of "dirty coal" to foul machinery. Third, both the amount of energy required and the need to cut unit costs demand that developers come up with coal conversion hardware that can successfully handle materials on a gigantic scale.

The United States today sits on one of the most plentiful, but up to now largely useless, forms of energy in the world—shale oil. While no precise data exist, the total reserves of shale oil in the United States are comparable with all other reserves of petroleum worldwide. In Colorado's portion of the Green River formation alone, estimates have ranged as high as 500 billion barrels. Oil exists in shale in the form of an organic substance called "kerogen." This substance can be turned into useful oil by a retorting process in which shale rock is heated to a temperature of around 900° F. While retorting techniques have been around for a long time, thus far, none of them is economically attractive.

Various industrial firms have been experimenting with prototype systems for extracting oil from shale with the hope of coming up with an economical process. Most of these firms are exploring the technology of retorting the shale above ground after the rock has been mined by conventional means. One company is taking a different approach. It is experimenting with methods for extracting oil from underground shale without mining it. Avoiding the need to mine huge amounts of rock should reduce costs considerably, as well as greatly lessen the environmental problems associated with the disposal of gigantic quantities of wastes. In this concept, holes are drilled into the shale, a room cavity is excavated, natural gas is forced into the cavity, and then burned or exploded. Cracks result in the rock and the shale is heated to a point that releases kerogen. Subsequently, the kerogen can be pumped to the surface and refined.

Major technical and environmental problems (including the enormous need of shale exploitation for water, scarce in the Western states where shale abounds) confront those committed to building a commercial shale oil industry. There is considerable debate whether the technical and environmental difficulties of any of the different approaches can be overcome within the foreseeable future. If they are, R&D will have to play a key role.

The heat contained within the earth itself offers a largely unused energy resource in the United States. We call this energy source "geothermal power." Some countries like Iceland and Italy already are tapping the steam and hot water that reaches the surface unaided for local heating and electric power generation. Moreover, the exploitation of this power source is relatively pollution free. The United States, it is believed, has major geothermal zones, especially in the West. The estimates for geothermal energy potential for electric power generation vary greatly. Estimates range from a low of 30 thousand megawatt-centuries to a high of 100 million. The difference occurs because of varying assumptions regarding geothermal conditions, state of the art, economics, and the impact of future R&D. Some researchers are convinced that geothermal power potentially can play a significant, if not a dramatic, role in helping to relieve power shortages.

Specifically, geothermal energy of potential commercial significance exists in the form of hydrothermal reservoirs containing steam, hot water, or both, in geopressured reservoirs, in near-surface intrusions of volcanic molten rock, and in impermeable hot rock crystalized from molten rock.

The state of the art of geothermal energy utilization technology limits us to the use of relatively clean, high-temperature steam and hot water systems. If the higher estimates for power generation are to be realized, we must develop advanced technologies. These would include deep and high-temperature drilling methods, reservoir engineering, production management and well stimulation techniques, improved turbines and power cycles. Advances in deep well pumping systems, mechanical and chemical control of mineral deposits, corrosion abatement, methods for disposal of waste fluids, and containment of noxious gas emissions would also help. Vigorous R&D programs are needed in all these areas if geothermal power is to reach its potential.

Some researchers are looking to the sun to meet mankind's long-range energy demands. In fact, so many alternative approaches abound that it is difficult to decide which are the most promising. The energy that reaches the earth each year is equivalent to burning 120 trillion tons of coal, about 25 times the world's estimated reserves. Economic exploitation, however, is hindered by two factors. First, the sun's energy is very diffuse; it exhibits a very low intensity, averaging about 108 watts per square foot of receiving surface. Therefore, the sun's rays must be concentrated into spaces small enough to produce sufficient, useful power. To do this requires a sunlight collector, composed of heat-absorbing surfaces. Second, there is a problem of supplying power when the sun is not shining. How is the sun's energy stored during periods of sunlight and released during periods when the sun doesn't shine? Yet, the potential of this enormous, pollution-free energy has prompted enthusiasts to advocate intensification of R&D activities in order to convert what are now experiments to commercially feasible energy systems.

There are two major approaches envisioned for exploiting solar energy: (1) generating electric power, and (2) tapping the sun's energy directly for space heating, cooling, and water heating.

While techniques for large-scale use of solar energy are known, the conversion costs are presently considered so high that this energy source cannot compete either with fossil fuels or nuclear power. This fact has not kept advocates of solar energy from advancing ideas and experimenting with designs. Modern solar thermal power plants rely on the development of a variety of technologies such as precision sun-following mirrors, better heat-absorbing surfaces, and plastic lenses.

Some researchers have suggested grandiose schemes for using solar energy for major power generation. One scientist has proposed the idea of huge satellite solar power stations, weighing some 25 million pounds and orbiting the earth at a distance of about 22,300 miles. He estimated that such a system could generate 5,000 megawatts, enough to meet the annual needs of 10 million people. Others have suggested the use of solar farms in which solar collectors would spread out over a total of 15,000 square miles in the sunny Southwest. These huge facilities could develop an electric capacity of 1 million megawatts, enough to satisfy the country's electric power requirements until the year 2000. Both these potential projects not only present major technical challenges, but also entail enormous costs. They seem destined for the remote future, if they are realized at all.

The more immediate application of solar energy relates to space heating, cooling, and hot water supply. In the United States some 1.5 billion tons of coal equivalent annually is used for space heating. This constitutes between one-fourth and one-fifth of total energy consumption. In some areas air conditioning uses more energy than space heating. To date a sizable number of experimental solar heated and cooled dwellings have been built in the United States. While each uses a somewhat different approach, in most cases their roofs are covered with flat glass or plastic collectors that trap the sun's heat underneath, much as a greenhouse does. The trapped thermal energy is then used to heat air or water, which in turn acts to heat the house. For use at night or on cloudy days the thermal energy is usually stored in insulated water tanks or in bins containing gravel or some other heat absorber. The technology of solar heating and cooling already exists and optimists contend that with improvements this technology soon could be used on a large scale.

Researchers at the University of Delaware are examining a system, designed to convert sunlight directly into electricity for space heating and cooling. Instead of using passive black surfaces, they employ solid state devices, solar cells, which produce electricity directly from the sun and do so without moving parts. Silicon solar cells have been used for some 15 years to power almost all of our space satellites. Silicon cells still are very expensive, but developers are hoping to cut their costs dramatically. Not quite as efficient, but much less expensive, are cadmium sulfide solar cells that convert solar energy into 50 watts of electric power per square meter of surface. The University of Delaware project has built a house in which the cadmium cell solar system is used experimentally.

In sum, if and when solar energy becomes a major source of power, in large part, depends on the effectiveness of the R&D effort made in this field. Some observers feel that semicommercial solar heating and cooling systems could be available within 2 or 3 years. The prospects for competitive solar electric power generation, however, are not bright in the short run, but perhaps could prove so in the distant future.

To a considerable extent, this nation has staked its energy future on the development of large-scale, economical, nuclear power generation. After some 30 years and investments of millions of dollars, we now have an economically attractive fission reactor technology (the principle of the original atomic bomb). In 1973 the nation had 30 atomic reactors in operation, 60 under construction, and 75 on order. With 150 years of operating experience in the United States, the federal government believes that the safety of nuclear power has been clearly proved (although some critics still suggest that it is unsafe). At present atomic power supplies about 4 percent of the nation's electricity. By 1985 its share is estimated to grow to some 25 percent and up to 60 percent by the end of the century. Thus, the current nuclear capacity of about 14.7 million kilowatts is expected to grow to 1.2 billion kilowatts by the year 2000.

President Nixon concluded that our best hope for meeting the country's growing energy demand lies with the fast breeder reactor. The normal type of fission reactor uses up all its uranium fuel. The breeder reactor uses the uranium, extracts the energy, and forms more fissionable products (plutonium), that are used as fuel in the same reactor. According to its developers, the breeder reactor could extend the life of our natural uranium fuel supply from decades to centuries, with far less impact on the environment than today's power plants. For several years, the Atomic Energy Commission has placed the highest priority on developing the liquid metal fast breeder. In 1972 the project was reported ready to move out of the laboratory and into the demonstration phase with a commercial size plant. Yet, major technical obstacles to constructing a demonstration plant of some 300 to 500 megawatts exist. Moreover, the environmental aspects of the fast breeder have been challenged. Nonetheless, the Administration still feels that the promise of the fast breeder is very great and has increased the funding to support the program.

Many exciting possibilities have been advanced for utilizing large-scale nuclear power reactors. For example, the coupling of atomic power with desalinization of sea water could bring rapid land reclamation and attenuate some of the adverse consequences of the world's population explosion.

For nearly two decades, nuclear scientists have undertaken a sizable research effort designed to harness the almost limitless energy of nuclear fusion (used in the nuclear bomb). Recent progress suggests the scientific feasibility of this project may be demonstrated in the 1970s. To increase the probability of success, the federal government has increased funding for this project. Yet, recognizing the difficulty of moving from the laboratory to engineering reality, nuclear scientists have cautioned against expecting the early widespread use of controlled fusion as a major source of electric power. Commercial fusion energy may not appear until the 21st century.

Researchers also have developed radioactive isotopes into a useful tool for man. These atomic products have proved useful as tracer elements in medical research and carbon-14 now helps date archaeological findings. With continued R&D a host of new applications for radioactive isotopes seems certain to emerge.

Physics

As a frontier discipline, physics not only unearths new fundamental knowledge, but points the way for the design engineer. For example, a single rocket system, in addition to mechanics, hydrodynamics, structural mechanics, and other disciplines, may also call upon the latest in microcircuitry that solid state electronics has to offer, while data from solar physics will provide much of the navigational know-how necessary for successful flight.

Pulsars, the recently discovered, rapidly pulsating radio sources observed in the sky, have attracted great interest. These phenomena appear to represent such an extreme form of behavior that what they do and are may be of fundamental importance of our understanding of the basic physical laws of the universe. Scientists have pointed out that pulsars would most likely exhibit phenomena such as superconductivity and superfluidity which are more normally associated with materials that are manipulated by solid state physicists.

Some operations of technological systems, such as communications, surveillance, guidance, and data processing have been transformed by solid state physics. Continued development of solids remains essential for a better understanding and use of materials in general, thus leading to improved devices and new applications. Optical crystals, new magnetics and superconductors, and photoconductive light detectors are some of these new materials. The fields of magnetics and semiconductors recently have come together and may have great technological importance in the areas of electronics, communications, and magnetometry. This synthesis involves the coupling of magnetic and electronics states to produce new materials and devices for electronics.

Research in low-temperature physics offers a promising future. At ultra-low temperatures materials become superconductors of electric current. Advances in superconductivity point the way to success in yielding improved devices having increased efficiencies such as accelerators, gyros, magnetometers, computer components, and microwave detectors. Emphasis is being placed on developing more efficient refrigeration units.

Research in plasma physics (ionized gases) could lead to major advances in several important fields. Certainly, if we want to fashion effective controlled nuclear fusion as a future alternative source of energy, we shall have to deepen our understanding of plasma phenomena. Continued investigations also could help us learn more about the impact of solar activity on communications and navigational systems. Scientists want to resolve questions relating to the dynamics of fully and partially ionized gases. Research in atomic and molecular physics will enhance what we know about plasma systems.

Atomic science, more than any other scientific discipline, comes closer to the heart of questions involving the fundamental nature of matter, energy, and the physical universe in which we live.

While we are well into the atomic age, many questions remain unanswered in nuclear physics, including the basic problem of what holds an atomic nucleus together. Another problem is that of classifying what scientists call "fundamental particles." Often we know little of the rhyme or reason for the existence and behavior of these fundamental particles. They leave their tracks on photographic plates when properly manipulated, and cast some doubt on our understanding of some of the laws of physics. For example, some scientists suspect that over extremely short distances light may travel faster than the speed postulated by earlier physicists.

New insight into the nature of fundamental particles and other aspects of nuclear physics may be gained by using more powerful high-energy particle accelerators.

New Materials

Significant progress has been made in certain fields in the materials area. Exciting developments have appeared in the field of composite materials that reinforce other materials. Composites have widening use in aerospace structural applications. In the past, Fiberglas-reinforced composites have proved especially useful. Advanced composites promise to meet today's even more exacting demands for improved toughness, ease of fabrication, resistance to very high temperatures, and better mechanical properties. In searching for composites of high strength, low density, and the degree of stiffness essential to structural applications of major components, researchers have successfully developed boron filaments that meet very rigid specifications (e.g., 400,000 psi[2] tensile strength).

Recently improved boron filamentary materials and promising high potential reinforcement materials, such as graphite, have led to significant cost reductions. Materials once costing $1,000 per pound have been cut to $100 per pound with an accompanying increase in strength (from 300,000 psi to 525,000 psi) and improved quality. Moreover, weight savings in structural materials have permitted gains in the performance of aerospace vehicles. We are now able to achieve advances in aerodynamics, flight control, and noise reduction with a minimum increase in weight. Boron composites reduced the weight of the FB 111 wing tip by 15 percent, the F-5 main gear door by 36 percent, and the Pratt and Whitney JFT 22 jet engine fan blades by as much as 50 percent.

Another benefit of composites is their increased reliability over metallic structural materials used previously. Thus, the military

---

[2] Pounds per square inch.

has expanded its use of composite materials in building the F-14 and F-15 fighters and the B-1 bomber as well as in the tail cone of the CH-54 heavy-lift crane helicopter. The use of these composites in space and civilian applications should grow.

Substantial progress also is being made in the technology of ceramics. Solid state physicists are investigating high temperature reactions and structures, but more work is needed to synthesize predetermined properties. Plastics and plasticizers also show promise, but thus far they lack careful product control. If plastic-oriented materials are to be used, the technology of these new materials must be improved to provide means of ensuring rigid quality control.

Rare earth additives promise to improve mechanical properties of steel and other metals. By introducing such rare earths as cerium, lanthanum, neodymium and praseodymium, metallurgists have improved the impact strength of steel plate as well as producing steel that has greater ductility and formability. These advantageous properties depend largely on the sulfur content that remains in the finished product. At certain strengths, sulfur adversely affects the composition and strength of finished steel. Thus, sulfur either must be removed or be changed to a form in which it is less harmful. The addition of rare earths can do either, depending upon the product need. This development should prove a significant advance in steel metallurgy.

Electronics

Perhaps no field of technology has experienced greater progress than that of electronics. While advances have been made in many areas of electronics, including making more effective use of the radio spectrum, two lines of development seem to offer great potential— lasers and microelectronics, and the discussion that follows will be limited to these two fields.

The laser (light amplification by stimulated emission of radiation) is a device that radiates and amplifies light waves in a narrow and extremely intense beam of light. Lasers may use either a gas or a solid material (such as a ruby crystal) for purposes of stimulated emission. In a laser, electromagnetic energy stimulates atoms in the gas or crystal in a way that releases light particles at a particular wavelength. The resulting light source has several unique characteristics: it offers an almost perfectly focused beam of light, with minimum loss by spreading; it consists of one color; it is coherent (the light waves are in phase with each other rather than emerging in random fashion as occurs with ordinary light sources); and it can be made enormously intense as well as be sharply directed.

Researchers have discovered a number of very attractive industrial, scientific, and medical applications. Probable fields of application include communications, optical radar systems, energy transmission in space, surgery, diamond cutting, welding, measurement, and a great many highly specialized uses in research, testing, and manufacturing.

The U.S. armed forces have been the chief customers of industry's laser development and production. In 1971 they spent some $80 million for laser range finders, bomb guidance, fuses, and prototype devices. Laser-equipped "smart bombs" proved very effective during the bombing campaign over North Vietnam in 1972. There has been speculation about a laser "death ray" that would make nuclear missiles and their defenses obsolete. At the present stage most authorities regard such projections as futuristic. Foreseen, however, are improved laser weapons for tactical use in bomber defense and air-to-air missiles.

Some analysts have estimated that laser sales for civilian uses could reach $500 million to $1 billion a year by 1980. Today, descendants of the original ruby laser are selling side by side with newer versions of solid lasers (chiefly glass and yttrium-aluminum-garnet) as well as the popular gas lasers (helium-neon, argon, nitrogen, and carbon dioxide).

The communications industry appears to be one of the chief beneficiaries of laser technology. The laser beam not only is free from outside interference, but its wide band width will carry thousands of times as much information as the wavelengths used for radio and television waves. Lasers can be hampered by weather conditions such as fog, rain, and snow, but research is seeking to overcome these difficulties. Moreover, these obstacles do not impede the use of lasers in outer space.

The semiconductor laser, a low-power, but highly efficient, extremely compact type of solid laser, is opening up exciting possibilities as a component in all kinds of electronic products. Still in the laboratory stage are liquid lasers that are tunable like radios and chemical reaction lasers that produce their own internal energy source. One communications laboratory has developed a laser smaller than a grain of sand that can be operated for up to a million hours on power produced by flashlight batteries. Researchers also are working to develop optical systems for space communications. Conceivably, hospitals, government agencies, and large companies could create a demand for laser communications to transmit private data on television hookups between two locations within a city.

Lasers have proved very useful as surveying systems for the construction industry. For example, commercially, the laser's most dramatic success thus far has been in the aligning of sewer pipes. With this tool, contractors can lay pipe 30 percent to 40 percent faster. In these alignment applications, the beam serves as an amazingly precise reference line. In addition to alignment uses, lasers increasingly will be applied to testing, measuring, cutting, welding, and drilling.

Through the use of lasers, scientists can create and interpret holograms—photographs in three dimensions that record the most minute detail in changes and imperfections in materials. Holograms are also being used to develop new ways to miniaturize information for computers, thus permitting dramatic savings in storage space. For example, an argon laser projects a book page—one of 2,500 stored on the 2-inch-square hologram film. This process is used to develop computer memories small enough for spaceships. Holographic techniques suggest applications in training devices, image quality improvement, and fabrication technology, as well as in data processing and display devices.

Medical applications could do much to improve the health of people. For some time physicians have been using lasers to spot weld loosened retinas of human eyes and to "vaporize" skin cancer. Scientists predict that if current research trends continue, the laser, because of its accuracy and quality of intensity, may be an important instrument in brain surgery.

The field of solid state technology has ushered in the age of microelectronics. We can now construct devices built of circuits so tiny that they literally can be balanced on the head of a pin. Electronics is largely concerned with the amplification of *radio* signals. The vacuum tube was the first amplifier to make radio feasible. A major breakthrough occurred in 1947 with the development of the transistor, a miniature amplifying device, made of germanium, that performs the functions of the vacuum tube in solid bodies. The transistor produces electrons for circuitry use by means of controlled impurities within crystalline substances. They are superior to vacuum tubes in that they are small, lighter, more reliable, more rugged, require less power, can perform electronic functions at extremely high speed, and cost considerably less.

As a result we have witnessed a fantastic growth in the use of transistor circuitry in military and commercial products. Because of its efficiency as a switching device it has enjoyed wide use in digital computers. Researchers have fabricated even smaller crystals which perform effectively as transistors as well as associated components of circuitry-resistors, capacitors, and diodes. One development is the integrated circuit, or chip, in which both the components and connections are imprinted within the crystal. Such a complete circuit, typically consisting of 10 to 20 transistors and 40 to 60 resistors, can be built into a bit of silicon measuring only about a tenth or twentieth of an inch on a side. Such "integrated" circuits can be manufactured simultaneously on a silicon wafer that is about an inch in diameter and less than one-hundredth of an inch thick. Great strides in micro-photoengraving have enabled us to "print" hundreds of miniscule circuits of these chips and wafers.

Microelectronics surely will bring about startling changes in many areas. Micromedical applications seem promising, from building ever tinier hearing aids, to microelectronic television cameras to explore the digestive tract, to connecting artificial limbs to nerve ends.

In modern aircraft and spacecraft avionics microelectronics is increasingly indispensable. Because of its small size and weight, microelectronic equipment enables vehicles to house more fuel and carry a higher payload. Equipment can be more complex and yield greater accuracy. Lighter weight and increased fuel allowance also permit significantly longer range. Moreover, the redundancy in systems that microelectronics makes possible both increases the reliability of hardware as well as cuts maintenance costs. As reliability increases, fewer vehicles are required to accomplish a given mission, thus lowering the costs of operation.

In the field of computers microelectronics can also make dramatic progress. Computer systems use circuits that are pulse driven, on-off operations, and are largely repetitive and low power. They need few capacitors and those employed are small in both electronic and physical size. Semiconductor integrated circuits prove ideal in these digital applications. The major challenge to microelectronics in digital systems is to devise a method for interconnecting the many functions of the computer system. Microelectronic components allow such interconnections in a smaller volume than otherwise possible. Someday we may see desk-model computers which would replace registers as well as inventory, payroll, and accounting machines. A superfast calculator using microelectronic circuitry has been developed costing under $100. This machine can calculate the most expedient, efficient ways to initiate loans, make time payments, invest in stocks and bonds, and gauge savings accounts and interest rates. It offers a simple, relatively error-free way to balance checkbooks, tabulate income tax, and manage household budget items. The new calculator performs all the functions of a standard adding machine (addition, subtraction, and multiplication) plus division; and the operator enjoys greater flexibility in using it. Its designers believe that within the forseeable future most homes will have such a calculator.

One of the most dramatic advances in the field of microelectronics is the transition of metal-oxide-semiconductor (MOS) technology from a laboratory curiosity into usable microelectronic devices. This technology has been widely accepted by commercial data processing equipment manufacturers for calculators, minicomputers, computer terminals, and other digital data processing devices (that provide for the input and output of data and perform basic arithmetic functions). This new microelectronic system reduces the number of circuits required for complex, professional calculators by up to half compared with present machines. Large-scale integrated circuits produced through the MOS process contain 500 or more transistors in a chip typically 0.1 inch square.

MOS circuits have been applied to a growing number of aerospace (including military), projects because the technology has all the characteristics important to avionics systems: compact size, low weight and power, high reliability, and low cost. Until recently, our limited experience with MOS, the speed limitations of early MOS designs, and a lack of reliability had hindered progress. Within the past few years, the shortcomings of early designs have been overcome and MOS technology has been successfully applied in several aerospace programs. MOS probably is now considered the leading candidate for long-life spaceborne computers that demand low power and freedom from failure on these long missions.

Another device that has profited from the MOS process has been a correlator that employs a pattern-matching technique for achieving missile terminal guidance, for aiding aircraft navigation, and for controlling industrial processes. For example, successive digitized aerial photographs can be compared, noting similarities and differences, a capability that has both military and civilian applications.

In addition to aerospace equipment, civilian-oriented calculators, credit and sales transaction terminals, cash registers, and business equipment have profited from the introduction of MOS large-scale integrated circuits. Using MOS technology, one manufacturer has developed automatic point-of-sale machines that promise to expand cashless purchasing by reducing the cost of designing and manufacturing credit card terminals.

Life Sciences

The biological sciences deal with all forms of plant and animal life and, increasingly, biological scientists have striven to provide quantitative descriptions of the way in which life systems operate. Such descriptions treat, at one extreme, the molecular structures characteristic of life and, at the other extreme, the interactions among different organisms and between organisms and their environment.

The biological sciences appear on the threshold of a revolution as remarkable as that which the physical sciences experienced by the discovery of radioactivity. The key terms in this biological revolution —DNA and RNA—still sound slightly esoteric, but are becoming very familiar. DNA stands for deoxyribonucleic acid, the giant molecule that carries the genetic blueprint in the nucleus of every living cell from bacteria to the most complex plant or animal. RNA, which stands for ribonucleic acid, is another giant molecule. It is produced by DNA and, in turn, directs the production of thousands of enzymes, organic catalytic agents, which trigger the metabolic machinery of the cell.

In the past 20 years scientists have come to understand something of the genetic role played by DNA and RNA, but much still remains to be discovered. It seems fairly clear that hundreds of thousands or even millions of molecular subunits combine to form a single DNA molecule. This process represents a "code" of instructions governing the transmission of characteristics from one generation to the next.

Scientists in biological research became excited about the possibility of "breaking the code" and by 1964, in large part, they succeeded in doing just that. This achievement has enhanced our understanding of biological processes and researchers now are working to refine this knowledge and discern its implications. It raises fascinating, even awesome, possibilities with regard to the synthesis or modification of DNA and, consequently, the control of genetic characteristics in plants and animals.

The revolution in biological science has tended, among other things, to dim the distinctions between biology and the physical sciences. In the molecular and submolecular fields, biological scientists are rapidly developing a common language with the physicists. At the same time, specialists in certain branches of the physical sciences, notably electronics, have begun to pay increasing attention to biological models which provide helpful insights into the design of nonbiological systems. A subspeciality of research, bionics, has emerged. This hybrid science draws upon the methods and knowledge of biologists, physicists, and engineers.

Scientists have achieved some successes in trying to understand the gene. Within recent years they have produced a single gene in experiments, including synthesizing a gene from elementary chemical units. In essence, these scientists exploited the chemical properties of DNA. It remains to be shown that these isolated genes can be reinserted into a cell and express their chemical information. Yet through experimentation it is possible that we will soon obtain considerable new biochemical information on how genes are regulated in living cells.

Substantial progress has been made in the last decade in developing small sensors and power supplies suitable for human and animal instrumentation. Current technology is based on miniaturization, specialized transducers, and power supplies with a high ratio of energy to volume and weight. It appears possible to produce bioinstrumentation systems far smaller than those now existing, which will be suitable for implantation in living tissue. With this equipment, biological researchers will be able to study energy expenditure of contracting muscles, chemical reactions taking place inside cells, and the transmission of impulses along various components of the nerve cell.

By acquiring new knowledge of the structures and mechanisms at work in relatively simple cells—e.g., certain bacteria, algae, and viruses—and by developing new investigatory techniques, scientists now can study directly and more effectively the far more complex cells of mammals, especially human cells. A systematic gathering of fundamental data on the structure, function, regulation, biochemistry, and genetics of normal and abnormal cells is making a major contribution to medical research. This line of inquiry has as its long-term objective learning to influence cell growth and functions for the benefit of man. For example, one researcher has demonstrated that the degree of acidity of the immediate environment of a cell has a profound effect on the cell's growth, on the cell population level, and on the behavior of the cell. Very small changes in the acidity of the growth medium have been found to affect both normal and cancerous mammalian cells. By manipulating the acidity, investigators hope to obtain important new insights into the nature of normal and cancerous tissues.

Biological research also is yielding new ways of controlling pest infestations that damage crops. Excessive long-term use of pesticides has produced insect strains that inherit resistances to pesticides. Biological investigators are experimenting with nonchemical ways of pest control. It has been discovered that sex attractants (secreted by insects) can be synthesized and used successfully as lures for insect traps. The use of hormones also shows promise. They have the potential for field use as disruptors of pest populations as they can dictate the growth and development phenomena of insects. Therefore, by feeding quantities of certain hormones to insects, it is possible to upset their normal growth pattern and prevent the insects from reaching the adult stage when they become harmful.

Increasing attention is being paid to investigations in tropical biology. The terrestrial tropics, because of man's impact, currently are undergoing enormous biological change. People throughout the world, and especially those living in temperate zones, are becoming more dependent on tropic regions of the earth for the resources that undergird modern industrial civilizations. But we need more knowledge of the tropics to increase the effectiveness of our resource exploitation and, just as important, to prevent ecological trauma. Tropical biotas and ecological systems are succumbing rapidly to man's invasion, but they are enormously complex, of great scientific interest, and as yet very poorly understood. Hence, biological scientists are devoting an increasing share of their resources to the study of tropical phenomena.

Environmental Sciences

Fields dealing with the physical environment have made substantial progress within recent years. In the atmospheric sciences, advances have occurred in understanding physical concepts of weather phenomena. Meteorologists are investigating methods of monitoring effluents on a national and global basis. The subfield of atmospheric chemistry has been devoting its attention to pollution-related research.

In the field of weather modification, scientists are trying to learn how to mitigate the destructive hailstorms of the Great Plains. They also are interested in learning more about warm fogs and warm clouds in general. Methods of producing precipitation from cold clouds and dissipating cold fogs are well known and many techniques are operational and in commercial use. The earth sciences are moving ahead rapidly in organic geochemistry, a relatively new field, and seismology (the science of earthquakes). Many aspects of the earth sciences are being pulled together by the unifying concepts of global tectonics—the theory of continental drift and seafloor spreading.

There is a growing awareness on the part of urban planners of the importance of input from the earth sciences in planning for the most beneficial and efficient use of land for our new cities. It is estimated that by the year 2000, tens of billions of dollars worth of new engineering structures will be built in areas of known earthquake activity. Knowledge of the local geology applied to the siting and engineering of these structures could result in reducing losses by up to 50 percent.

New experimental techniques are becoming available that permit subjecting materials to high temperatures at pressures of up to about 1,000 tons per square inch. These temperatures and pressures are equivalent to those deep in the earth's crust which cause the metamorphosis of minerals to various crystalline rock formations. The technique has been developed for geological experimentation, but it should have profound impact on research and on the possible synthesis of new mineral-like materials of exceptional properties.

Atmospheric physics has been of prime interest for decades because of its importance in weather forecasting and communications. Artificial satellites now permit the atmosphere to be observed continuously from above as well as from below, and a substantial increase in effort is required to properly utilize this new capability. Remote sensing also is offering new opportunities to understand the enviromnent. New sensor equipment can supplement the aerial camera to gather environmental information. They offer the promise of improvements in obtaining weather information. It also appears that new sensor systems may also be able to detect wave height, ocean surface temperature, and shipping traffic under all weather conditions.

Behavioral and Social Sciences

The life sciences, taken as a whole, include various behavioral disciplines which range from psychology to anthropology. Behavioral sciences also overlap social sciences, and some consider them part of the social sciences. These behavioral disciplines attempt to apply methods of experiment and observation, in the manner of the physical sciences, to the study of human actions within society. Consequently, much emphasis is placed upon quantifying data and using mathematical techniques.

The behavioral and social disciplines have not yet experienced a revolution matching that in the physical or biological sciences. Certainly, significant and challenging frontiers exist.

In many instances the practical effect of breakthroughs in the physical and biological sciences depends upon social and behavioral factors. By way of illustration, it appears that often science and technology change at a faster rate than cultural institutions. At the same time, social-cultural environments necessarily help condition the overall rate of change. In particular, they influence the willingness of societies to adopt modern science and technology and accept consequent changes in their social conduct and structure. The strain between old cultural ways and modern technology plagues emerging nations that desire to advance beyond their preindustrial, and in some cases, tribal past. These same tensions operate in different ways within industrialized nations. Behavioral and social sciences can give man an improved insight into problems of cultural adjustment.

The continuous refinement of knowledge of man's information-processing, problem-solving, and decision-making capacity under conditions of stress will come from research in the field of engineering and physiological psychology. Even though automatic and computer-controlled equipments are relieving men of many routine functions, the operation of these equipments requires highly specialized skills that are physically and intellectually demanding. Unless these demands become thoroughly understood, better integration of man and computer will not be possible.

Researchers are investigating the psychological factors that affect user acceptance of equipment. Mathematicians, operations analysts, and psychologists will have to cooperate to develop proper equipment acceptance procedures and to devise training programs that will allow men to operate complex technological systems to full potential.

As modern technological systems became more complex and more dependent on automatic control, the role and the performance characteristics of the operator must change. These new developments in equipment and equipment control point to an urgent need for studies of the man/machine relationship.

One of the most difficult problems in the man/computer relationship is the display of large quantities of information in a meaningful manner. It has become possible to assemble in a computer data from many sources and to present these data in virtually any type of display, in real time. Much more research is required to extract the maximum usefulness from this new electronic capability. In many cases it will require a new conceptual framework for attacking problems and making decisions.

The behavioral and social sciences have experienced progress, partly as a result of a methodological revolution. Until a very few years ago, behavioral scientists designed their experiments so as to minimize the number and complexity of calculations that they would have to perform—always a massive and time-consuming job where large data bases are involved. However, with the advent of the computer and the availability of relatively cheap computational ability, they no longer need avoid the massive computational loads and much of the field is rapidly becoming more quantified in its outlook and methodology.

Recent public interest has drawn attention to the study of social indicators—strategic and identifiable measures which indicate significant social changes. Replication studies in which sociological research is repeated after 10 or 15 years on a subject (such as the relative prestige of certain job categories, or the influence of education on income), make it possible to determine if there has been any basic change. Many studies now being conducted for the first time are being designed so as to permit exact replication in the future. A larger number and variety of such measures would supplement the base of our current understanding of change in America, which we get from such data as the census and economic indicators, and would enhance our knowledge of social phenomena.

Increasingly, problem-focused research draws its personnel from a variety of disciplines, and traditional disciplinary work has broadened to accommodate new problems. This is particularly true of urban problems and of foreign area studies. Emphasis of research in geography is now involved with urban planning, with the spatial distribution of income, with community locational decision, and with the perception of neighborhood by individuals.

Interdisciplinary Research into Social Problems

In recent years support has increased for scientific research on complex societal issues that require the contributions of diverse scientific disciplines. Scientists are investigating the potential for genetic mutations in man resulting from the introduction of man-made chemicals into the environment. They are developing several approaches for mass screening of mutations in man. They also are trying to fashion techniques for overall ecological evaluations, and they are designing computer simulation models to shed light on the cost and effectiveness of alternative environmental policies.

Drawing on the disciplines of engineering, applied mathematics, economics, political science, and city and regional planning, universities are conducting research on the technical, economic, social, and political aspects of environmental quality problems. Major aims include improvement in the framework for analysis of urban environmental problems and the training of graduate and postdoctoral fellows in conducting interdisciplinary research. Such study includes analyses of environmental management institutions, economic analysis of the household as an individual decision unit, and studies of municipal financing of environmental control systems.

Another effort has been directed at research into problems of solid wastes and fire protection. Data obtained from existing literature and a case study of the sources and disposal of solid wastes are being used to construct a benefit-cost model of the waste disposal system. Operations analysis techniques have been employed to deal with problems of fire protection, using computer simulation to develop estimates of benefits and costs associated with alternative protection systems.

Summary

While one cannot guarantee that all the aforementioned scientific and technological potentialities will materialize, more than likely, some of them will. Just one or two key discoveries or inventions could radically alter the way we think about our universe and lead our lives. In the past, the physical sciences have progressed more rapidly than the biological or social sciences. Some feel that the next dramatic scientific revolution may result from biological research. All that is certain is that in the coming decades the world will witness some startling advances, and, if history is any guide, some will take place in unexpected ways.

5 Science and the Federal Government

Krishnan D. Mathur and *Ralph Sanders*

It is clear that the federal government now plays a major role in supporting and guiding this country's science and technology, in devising science policy, and in managing research and development.[1] As will be seen, the founding fathers encouraged science and technological development. Through the years, the nation has created scientific institutions to help it discharge its duties in protecting the health and safety of its citizens and in promoting economic progress.

Yet, it seems that external threat more than domestic need triggered the massive influx of federal money and men into science and technological development. Two world wars and a 20-year period of acute international tension and competition catapulted the federal government into large-scale scientific activity. Growing Soviet military might and space capabilities presented the United States with a formidable challenge which affected its very survival. World conditions may change and with them the motivation for government support of and involvement in science and technology. Within recent years domestic societal problems increasingly have challenged the nation's scientific and technological capabilities. A Rubicon has been crossed, and science and technology will continue to remain a vital concern of the federal government.

---

[1] Literature treating the relations between science and government has been growing steadily since World War II. The following will prove useful to the reader: Vannevar Bush, *Science: The Endless Frontier* (Washington, D.C.: Public Affairs Press, 1946); Joseph S. Dupre and Sanford A. Lakoff, *Science and the Nation: Policy and Politics* (Englewood Cliffs, N.J.: Prentice-Hall, 1962); A. Hunter Duprée, *Science in the Federal Government* (Cambridge, Mass.: Harvard University Press, 1951); Robert Gilpin and Christopher Wright (eds.), *Scientists and National Policy-Making* (New York: Columbia University Press, 1964); Daniel S. Greenberg, *The Politics of Pure Research* (New York: New American Library, 1969); Don K. Price, *Government and Science* (New York: New York University Press, 1954); Don K. Price, *The Scientific Estate* (Cambridge, Mass.: The Belknap Press of Harvard University Press, 1965); and Jerome Wiesner, *Where Science and Politics Meet* (New York: McGraw-Hill, 1965).

Evaluation of Science-Government Relations

"I can not forbear intimating to you," observed President Washington, "that the expediency of giving effectual encouragement as well to the introduction of new and useful invention from abroad as to the exertions of skill and genius in producing them at home. . . . Nor am I less persuaded that you will agree with me in opinion that there is nothing which can better deserve your patronage than the promotion of science and literature." [2]

In this first annual address to the Congress on January 8, 1790, the first President of the United States echoed the sentiments of the American people. Even the early settlers had been conscious of the technological advancement of Europe and cognizant of the importance of science to the growth of a new nation. Among the *Mayflower* pioneers were skilled artisans, carpenters, and husbandmen, who brought with them tools, implements, and utensils, as early as 1620. Slowly the settlers designed a water wheel, invented a gearing mechanism, and constructed mills for softening wool.

By the time of independence, America had acquired considerable industrial skills. The colonists had mastered techniques in mining, metallurgy, printing and publishing, and construction. Benjamin Franklin's discoveries in electricity, Thomas Jefferson's experimental farms at Monticello, Benjamin Rush's pioneering work in sanitation and purification of water, David Rittenhouse's discovery of a comet in 1793, and Eli Whitney's invention of the cotton gin in 1798 attest to the growing spirit of American scientific adventure. Impressed by the American scientific progress, Joseph Priestly (the discoverer of oxygen) observed in 1794: "This (American) Republican government, by encouraging all kinds of talents is far more favorable toward the sciences and the arts than any monarchical government has ever been. . . . A free people will in due time produce anything useful to mankind." [3]

Our founding fathers felt that government should foster science and technology. President Washington sought to establish a national

---

[2] James D. Richardson, *A Compilation of the Messages and Papers of the Presidents,* Vol. I (New York: Bureau of National Literature, 1897), p. 58.

[3] Quoted in John W. Oliver, *History of American Technology* (New York: Ronald Press, 1956), p. 5.

center for the advancement of science. While differing on means, both Hamilton and Jefferson advocated the promotion of science. James Madison believed that Congress should establish a university for the diffusion of knowledge.

Both the states and the federal government aided scientific activities. After independence, the states retained freedom in scientific pursuits, sponsoring agricultural fairs, assisting industrialists in obtaining modern machinery, and advising farmers of new agricultural methods. They also undertook some major projects such as the building of the turnpike between Philadelphia and Lancaster in 1792 and the Erie Canal in 1825. The first state board of health was established in Massachusetts in 1869.

The major areas of state research are agriculture, health, education, and public works, all of primary concern to state governments. Originating as agricultural and mechanical colleges, state universities have made very significant contributions to science and technology. State research programs tend to emphasize life sciences rather than physical sciences. While state sponsorship of research and development does not nearly match that of the federal government, it is sizable and increasing.

Federal scientific leadership arose because state governments did not have the resources to make heavy financial commitments for research on a scale demanded by modern conditions. Nonetheless, the states still retain considerable scientific activities, which often are augmented by federal aid. The role of the states in science and technology has become complementary to that of the federal effort.

The federal government has emerged as the major sponsor of science and technology, although it is less dominant now than it was during the 1950s and 1960s. Federal responsibilities for promoting scientific activities evolved gradually in response to the needs of the times. After independence, in fulfilling its regulatory function, Congress enacted a patent law in 1790, provided for a decennial census, and instituted a uniform system of weights and measures. Not until the mid-19th century did the federal government become increasingly active in scientific pursuits. It began to grant funds for scientific enterprises, initiate investigations, and establish scientific organizations. In order to find the cause of accidents due to steam boiler explosions, the Treasury Department in 1832 granted funds to the

Franklin Institute of Philadelphia to undertake an investigation. This was the first federal contract to a scientific institution and resulted in the estabilshment of the Steamboat Inspection Service, the first scientific regulatory agency.

Persistent demands by farmers resulted in the enactment of one of the major legislative acts affecting science in this country—the Morrill Act of 1862. This Act created the Department of Agriculture and established the land-grant college, an institution which promoted cooperation in science and education between the federal and state governments. In the following years a host of agencies with scientific activities emerged: the Coast and Geodetic Survey, Weather Bureau, Bureau of Standards, and others. The federal government also supported and used the services of several quasi-governmental scientific institutions such as the National Academy of Sciences, founded in 1863.

World War I increased the scientific activities of the federal government. It sought to harness the nation's scientific and industrial resources to meet wartime needs. The federal government created several major consultative organizations—the Council of National Defense and the Naval Consulting Board (under the chairmanship of Thomas A. Edison). To bring the best scientific talent into war service, the National Academy of Sciences organized the National Research Council. The nation developed a system of consultation and cooperation on scientific matters among governmental agencies, industry, foundations, and universities. In addition, the National Advisory Committee for Aeronautics (NACA), created in 1915, contributed greatly to this country's progress in aviation.

Strong government support of scientific enterprise did not continue in the post-World War I years. The baneful effects of the war, the universal desire for disarmament, the stock market crash, and the depression of the 1930's dried up many sources for federal support of scientific efforts.

Not until World War II did governmental scientific activities burgeon again. The National Defense Research Committee (NDRC), created in 1940, undertook the weapons coordination task. High government officials soon saw the need for even greater cooperation between government, industry, and universities. As a result, President Franklin D. Roosevelt in June 1941 created the Office of Scientific

Research and Development (OSRD) under the chairmanship of a noted scientist, Vannevar Bush. OSRD became the top science authority and operating agency. It controlled funds, initiated promising scientific and medical research, carried developments to the stage of operating models, and let contracts with industry and universities. The military departments, however, remained the final judges of scientific developments having military uses. While OSRD was part of the Executive Office of the President, Dr. Bush had direct access to the Chief Executive.

Post-World War II

The ever-increasing problems of international tensions and Communist aggression after World War II necessitated greatly enlarged federal scientific activities for national security. In addition to national security, farsighted U.S. leaders saw the need to stimulate scientific and engineering efforts in order to improve the economy and general welfare. In a monumental report, *Science: The Endless Frontier*, Dr. Bush recommended that a permanent agency be established to support basic research. The result was the creation of the National Science Foundation (NSF). In another institutional milestone, Congress, after lengthy debate, established the Atomic Energy Commission to promote the development of atomic energy for both military and civilian uses. Military organizations also realized the value of promoting basic research which might not have an immediate military application. In 1946, the Office of Naval Research was established. After the Soviets launched Sputnik, the United States in 1958 set up the National Aeronautics and Space Administration (NASA) to spearhead accelerated U.S. explorations in space.

Since 1945, the whole gamut of government science activities has grown considerably. The structure of scientific administration and research became a complex web of governmental agencies and academic and private organizations acting under varied systems of contracts, grants, and in-house programs. To avoid unnecessary duplication of research, to make maximum use of scientific talent, to utilize the assets of new discoveries, and to meet the exigencies of a rapidly changing technology, some general policy guidelines had to be fashioned and all federal programs had to be well coordinated and constantly reviewed and improved. The executive branch then faced the difficult task of devising mechanisms for coordination, on the one hand, and for planning overall scientific activities, on the other.

Government increasingly sought high level technical advice in the formulation of national and agency policies and programs. Scientists moved into high offices, new technical divisions arose to improve and expand scientific programs, and politicans and administrators solicited the advice of specialists on various questions. The White House remained the controlling authority on all federal scientific programs and, in contrast to the past, the President increasingly concerned himself with scientific matters.

In 1947, President Harry S Truman dissolved the OSRD and replaced it with the Scientific Research Board under the Director of War Mobilization and Reconversion. Its chairman, John R. Steelman, prepared a report for the President recommending certain changes in the organization and administration of the nation's scientific activities, including the creation of an Interdepartmental Committee on Scientific Research and Development and a science adviser on the White House staff. President Truman in 1947 created the Interdepartmental Committee to improve the employment of R&D for national welfare and to encourage agency collaboration, but he did not establish the presidential staff science adviser position. In 1951 the Science Advisory Committee, composed of prominent scientists and engineers from both government and civilian scientific communities, was organized within the Office of Defense Mobilization to advise the President on national security aspects of science and technology.

The Soviet's Sputnik achievement in 1957 prompted President Eisenhower to place on his personal staff a Special Assistant for Science and Technology, marking the first time in the post-World War II period that an adviser for science had entered the innermost circles of policy determination. A few weeks later, President Eisenhower again moved to strengthen his scientific advice by relocating the ODM Science Advisory Committee (SAC) within the White House as the President's Science Advisory Committee (PSAC). This committee was a semiautonomous group, specifically shielded from Congress, and composed of 17 eminent nongovernment scientists and engineers with the President's Special Assistant as chairman. Even more than the SAC before it, PSAC made extensive use of panels to study specific problems, thereby giving both government and private scientists the opportunity to contribute to policy formulation.

The PSAC, among its first actions, proposed establishing a counterpart group of policymaking representatives from government agencies engaged in research and development. The Committee saw the need for greater coordination of the federal R&D effort and, in effect, admitted that the Interdepartmental Committee had been only partially successful. As a result, the President in 1959 founded the Federal Council for Science and Technology (FCST) within the White House and eliminated the Interdepartmental Committee.

The Federal Council, chaired by the Special Assistant for Science and Technology, had 11 members, all individuals highly placed in the research and development program of those agencies most concerned with R&D—Defense, Agriculture, Commerce, Interior, HEW, AEC, NSF, State, Transportation, NASA, and HUD. The Council has been described as "essentially a sub-Cabinet for Science," whose members have "both a political or policy position and a technical position."[4]

Inasmuch as the Council was organized in committees composed of scientist-administrators, these committees, in turn, utilized panels of scientists from all federal agencies interested in a committee's mission. A considerable amount of scientific and technical talent was thus brought to bear on policy matters at least on a part-time basis. While this mechanism improved governmentwide consideration of science and technology, the prime responsibility for planning and implementing R&D programs remained with the departments and agencies; the Council used mostly mediation and persuasion and avoided directives.

Another important organizational change took place in June 1962 when President John F. Kennedy instituted the Office of Science and Technology (OST) to provide him with staff support. OST, staffed by some 20 full-time professionals, examined major policies, plans, and programs and their relationships to the general welfare, national security, and foreign policies; evaluated scientific and technological advances in relation to their effect on national problems; considered the effect of federal programs on nonfederal resources and institutions; and investigated methods to strengthen the scientific and engineering communities. It provided, for the first time, formal communications

---

[4] See testimony of Dr. Jerome Wiesner, former scientific adviser to the President, in U.S. Congress, House, Committee on Science and Astronautics, *Government and Science, Hearings before the Subcommittee on Science, Research and Development*, 88th Cong., 1st sess. (Washington, D.C.: U.S. Government Printing Office, 1964), p. 173.

on science matters between the White House and Congress. The OST also supplied staff support for PSAC and FCST, a support which in the past had been rendered by the rather small organization of the President's science adviser. While OST's power remained rather limited, it did contribute a "selective mechanism for dealing with policy problems arising out of research and development, and for providing a framework of coordination which stops short of interference with the immediate responsibilities of departments and agencies."[5]

During President Nixon's first Administration the science policy machinery of the White House continued to provide the Office of Management and Budget with technical advice in fashioning the chief tool for allocating resources within government. Although the growth of science and technology had increased the difficulty and complexity of budget making, the OMB resisted hiring scientists and engineers to review R&D proposals. It preferred to use a broad approach in examining agency requests from the standpoints of public policy, availability of money and manpower, and balance among alternative programs. It relied upon other organizations, especially OST, for sophisticated judgments of the science content and technological integrity of programs. The Domestic Council, a high-level White House group established by President Nixon, is concerned with how science programs fit into overall domestic policy. On special issues, consultation takes place with interested organizations, such as the Council on Environmental Quality on pollution topics.

Some people have suggested the creation of a Department of Science. Such an agency would enjoy Cabinet status and would centrally administer the scientific activities of the federal government, thereby permitting more national long-range planning and avoiding unnecessary duplication. This is not a new idea. In 1884 a congressional commission considered amalgamating certain scientific agencies, such as the Geological Survey, the Coast and Geodetic Survey, and the Hydrographic Office. The commission concluded that unification was unnecessary. In 1961 a congressional committee, chaired by Senator Henry M. Jackson, also concluded that since all scientific

---

[5] See testimony of Dr. Jerome Wiesner, in U.S. Congress, House, Select Committee on Government Research, *Federal Research and Development Programs,* (Part I, 88th Cong., 1st sess. (Washington, D.C.: U.S. Government Printing Office, 1964), p. 566.

activities could not be centralized effectively, there was no need for a Department of Science. While many advocates of a Cabinet department remain vocal, more observers agree that "Science appears in so many different places that to create a department to include only a few agencies would not help, but hinder the inter-relatedness of science."[6]

Federal R&D Institutions

Federal activity in science and technology is spread across nearly every department and agency. For the most part, agencies either support R&D to further their assigned missions (e.g., Atomic Energy Commission) or provide grants and contracts to stimulate scientific investigation of a more general nature (e.g., National Science Foundation). From a national point of view, the basic problem in federal administration of R&D is that the various scientific disciplines and technologies do not coincide with agency missions.

Under the U.S. governmental system, departments and agencies have the responsibility of initiating and implementing programs, including selecting competitive proposals, as well as monitoring contracts and grants. Where possible, the presidential science bodies tried to formulate national programs. The case of oceanographic research illustrates this point. For a long time federal oceanographic activities were conducted in a disjointed way by some 19 agencies. In 1960, at the suggestion of PSAC and the National Academy of Sciences to expand government efforts in oceanography, the Federal Council organized the Interagency Committee on Oceanography (ICO). The members of the committee represented eight major departments and agencies having an interest in this line of research. The most significant accomplishment of the ICO was the development of a long-range National Plan which serves as a framework within which the specific program for each agency is formulated. Since 1966, the ICO has undergone several reorganizations. At the end of President Nixon's first term, it had become the Interagency Committee on Marine Science and Engineering within the Federal Council. The agency representatives on this committee occupied higher-placed positions than on the old ICO. Moreover, the new committee carried greater weight than did its predecessor.

---

[6] James L. McCamy, *Science and Public Administration* (Montgomery: University of Alabama Press, 1960), p. 104.

As indicated in Chapter 2, the Department of Defense conducts the single largest R&D effort in the federal establishment. It seeks to tap the fruits of scientific and technological investigations in order to fashion superior weapons and supporting equipment for use by the military forces of the United States. While the Department of Defense maintains over 130 laboratories of its own and utilizes the services of nonprofit institutions and university research centers, by far most of its research and development is carried on by private industry. In large-scale development work, the Department usually contracts with large firms, and these "prime contractors" then subcontract portions of the task to smaller, more highly specialized companies. In the aerospace industry the large corporation has become the conerstone of the defense technical base.

Outside of the Department of Defense, the Atomic Energy Commission (AEC)[7] and the National Aeronautics and Space Administration (NASA) carry on the most extensive scientific and engineering programs. The AEC is responsible for developing, using, and controlling atomic energy to advance the security, economy, and general welfare of this nation. As in the case of the Department of Defense, the operations of the Commission are carried out largely by industrial concerns and by public and private institutions under contract. The principal production and research and development activities are conducted by contractors in facilities owned by the AEC.

The U.S. space program is administered by NASA. It undertakes R&D programs for the solution of problems of flight within and outside the earth's atmosphere and develops aeronautical and space vehicles. In conducting activities required for the exploration of space, it arranges for effective utilization of U.S. resources, cooperates with other nations, and disseminates the results of space missions. In promoting both manned and unmanned space flights, NASA owns and manages a number of field installations. The bulk of its funds, however, are spent in securing the services of R&D contractors, purchasing equipment, and performing related management functions.

---

[7] In October of 1974 Congress passed a law dividing the AEC into two separate agencies. The Energy Research and Development Administration is charged with developing technologies to give the United States the capability to achieve energy self-sufficiency. The Nuclear Regulatory Commission is charged with ensuring the safety and security of the nuclear industry and other radioactive materials. This change is scheduled to go into effect in early 1975.

The Department of Agriculture, one of the oldest science-oriented agencies, has developed a number of semiautonomous research institutions permanently established in every state. Two major scientifically oriented agencies of the Department of Commerce are the National Bureau of Standards (NBS) and the National Oceanic and Atmospheric Administration (NOAA). The National Bureau of Standards, established in 1901, is the chief agency for setting codes and specifications and in developing and using an approved system of physical standards and measurements. NOAA was created in 1970 to improve man's comprehension and uses of the physical environment and its oceanic life. Its formation brought together the functions of the old Environmental Science Services Administration (and its major elements: the Weather Bureau, Coast and Geodetic Survey, Environmental Data Service, National Environmental Satellite Center and Research Laboratories) and other fishery and oceanographic organizations of other departments.

There are three major divisions within the Department of Interior with scientific interests. The Geological Survey conducts surveys and investigations in broad areas of topography, geology, and mineral and water resources. The Bureau of Mines conducts research to improve the productivity and safety of the nation's mineral-exploitation enterprises. The Bureau of Reclamation conducts investigations associated with conservation, development, and use of water, especially in the more arid regions of the Western United States.

The major scientific activities of the Department of Health, Education and Welfare are conducted in the fields of health. Research is primarily aimed at the diagnosis, treatment, control, and prevention of physical and mental diseases of man. The National Institutes of Health (NIH) conducts and sponsors most of this research. The NIH is composed of a number of institutes, each devoted to specialized research, e.g., cancer, heart, and others. The Public Health Service conducts and supports research and training in medical services while the Food and Drug Administration undertakes research designed to protect the public from adulteration and contamination of food, drugs, and related products.

The fundamental purpose of the National Science Foundation is to strengthen basic research and education in the United States. Since its creation in 1950, it has sought to accomplish this mission by providing contracts, grants, and facilities for research and for expanding and upgrading scientific education. It also assists schools and colleges in purchasing scientific equipment, participates in international scientific conferences, exchanges and disseminates scientific information, and provides indexing, abstracting, and translating services.

Under President Nixon the NSF has given great emphasis to its newly formulated Research Applied to National Needs (RANN) programs. The purpose of RANN is to focus scientific research on selected societal problems of national importance with the objective of contributing to the knowledge required for their practical solution. RANN is intended to bridge the gap between the basic research programs supported by the NSF and the research, development, and operations of the mission agencies. RANN chiefly uses universities to implement its projects.

Several regulatory agencies such as the Federal Aviation Agency, the Federal Communications Commission, the Federal Power Commission, and the Interstate Commerce Commission carry on scientific programs designed to help them achieve their missions of regulating air traffic, telephone and telegraph services, energy systems, and interstate transportation, respectively.

The Smithsonian Institution was established by Congress in 1846 under the terms of the will of a wealthy Britisher, James Smithson. In 1829 Smithson bequeathed his fortune to the United States to found an establishment devoted "to increase and diffusion of knowledge among men." The Institution has played a major role in the advancement of science in this country. For example, it financed Dr. Robert H. Goddard's early pioneer research in rocketry and recommended the formation of the Weather Bureau. It conducts basic research in physical and biological sciences; carries on technological, historical, and cultural investigations; manages art galleries, zoological parks, and museums; and maintains an International Exchange Service for disseminating scientific information.

The National Academy of Sciences serves as a quasi-governmental organization of distinguished scientists, devoted to the furtherance of science and its use for the general welfare. Although not a governmental agency, the Academy has enjoyed close relations with the federal government. Its great contributions have been in interpreting broad principles of science for the good of the nation and in providing the government with advice on science matters.

The National Research Council, organized in 1916 to aid the war effort, is part of the Academy and both operate under a unified governing board. The president of the Academy acts as ex-officio chairman of the Council. Neither of these distinguished bodies receives appropriations from the government. However, major financial support comes from contracts and grants from federal and state agencies in addition to funds received from private organizations, especially foundations.

A National Academy of Engineering, devoted to applying the resources of the engineering profession to the great technological problems facing this nation, was established under the charter of the National Academy of Sciences in 1964.

Major Changes under President Nixon

President Nixon introduced several major innovations both in the direction of science and in the way the federal government makes its science policy. First, he emphasized linking the nation's science policy and performance to the problems afflicting the U.S. world trade position and domestic situation. During his first four years in office he took several steps in this direction. In appointing Dr. Edward E. David, Jr., a research director of Bell Telephone Laboratories, the President, for the first time, chose a science adviser who primarily had an industry background. Industrial scientists increasingly began to fill positions within the higher councils of the advisory apparatus, previously dominated by university scientists. He also created a special consultant to focus efforts designed to expand the nation's technological capacity.

Second, for the first time in history, on March 16, 1972, the President sent Congress a special message on science and technology. This message outlined the science and technology needs of the country and the Administration's response to those needs. It cited the requirement to strengthen the federal role. In addition to increasing his 1973 budget request 9 percent over that of 1972, the President moved to develop an overall approach to the allocation of federal scientific and technological resources. However, he did not institute a yearly state of science and technology report like the reports of the Council of Economic Advisers.

Even before the special message of 1972, President Nixon sought to harness science for practical purposes. In 1971 he directed his

Domestic Council to examine new technology opportunities in relation to domestic problems. President Nixon reoriented the space program to focus on domestic needs (e.g., communications, natural resources exploitation) and moved to set certain civilian targets. In the special message he identified a number of areas where new efforts appeared most likely to produce significant progress, including (1) providing new sources of energy without pollution; (2) developing fast, safe, pollution-free transportation; (3) working to reduce the loss of life and property from natural disasters; (4) improving drug abuse rehabilitation programs and efforts to curb drug trafficking; and (5) increasing biomedical research efforts, especially those concerning cancer and heart disease.

In his message, the President likewise proposed ideas to strengthen R&D in the private sector chiefly through cost-sharing agreements, procurement policies, or other arrangements. To improve the climate for innovation the President took steps to encourage bridges or coupling arrangements among the various components of the nation's R&D enterprise: universities, industry, government laboratories, non-profit institutes, and state and local governments. He also saw the need for government to transfer the results of its R&D to wider use in the public sector. Thus, in 1970 he created the National Technical Information Service in the Department of Commerce and approved a change in the government patent policy which liberalized the private use of government-owned patents.

At the outset of his second term, President Nixon instituted a major governmental reorganization that had a dramatic impact on the White House's science policy machinery. On January 5, 1973, he announced a sharp reduction in the overall size of the Executive Office with the stated objective of reorienting that Office back to its basic purpose of assisting the President in top level policy and management matters. As part of this reorganization, the President abolished the post of his science adviser and the Office of Science and Technology, transferring their functions to the National Science Foundation and its Director.

In setting forth the reorganization plan, the White House said, "With a growing range of capability in the National Science Foundation, the President will now look to its Director as a principal adviser in science and technology matters." In his new capacity as science adviser, the Director assumed the task of advising and assisting the

White House, the Office of Management and Budget, the Domestic Council, and other entities within the Executive Office on matters where scientific and technological expertise is called for. He also was charged with acting as the President's representative in selected cooperative programs in international scientific affairs, including chairing such joint bodies as the U.S.-U.S.S.R. Joint Commission on Scientific and Technological Cooperation.

With respect to civilian research and development, the Director was charged with appraising the overall effectiveness of ongoing federal efforts and assumed the responsibility of recommending policy and program actions to achieve national goals.

The Director has several organizational relationships to support him in his activities. First, the President designated him as the chairman of the Federal Council for Science and Technology, a duty he already had undertaken in an acting capacity. As such, the Director has access to the views of major figures in the research and development areas of many agencies as well as an opportunity to help coordinate their programs. Second, he created within the NSF two offices to assist him in his role as science adviser. The Science and Technology Policy Office was designed to assist him in choosing priority problems for analysis and for coordinating the resources necessary to provide in-depth review of selected issues. In large part, this office performs the functions of the old OST. The Energy R&D Task Force was established to prepare background studies analyzing energy R&D objectives, needs, and priorities. It develops a report which presents detailed recommendations to an overall energy R&D report prepared by the FCST.

No formal replacement has been created for the defunct PSAC. The Director instead preferred ad hoc arrangements with qualified scientists and technologists from the academic, industrial, and research institute communities as well as from professional societies. He expressed the hope that the many outstanding individuals who in the past had contributed to the solution of science and technology problems would continue to render their advice and services.

In the case of national security, the President indicated that he would continue to draw primarily on the Department of Defense for advice regarding military technology.

The wisdom of the decision to remove the science advisory machinery from the White House triggered some intensive debate. Critics charged that the loss by science and technology of a channel to the White House was a serious one. Moreover, the coordinating and ambassadorial roles of the President's science adviser, especially in international science and technology, could not easily be filled by other institutions. Above all, they complained that only the White House could give the proper backing to the considerable study and overview effort needed to get an adequate understanding of the federal research and development enterprise. Without an OST and PSAC, penetrating investigations became exceedingly difficult.

Critics likewise questioned the effectiveness of the alternative advisory institutions set up. They doubted that the Director of the NSF had enough influence with the White House to ensure that the voice of science and technology would be heard or heeded. They questioned whether he could really help resolve vital, controversial, and complex issues. The Director, according to the critics, suffers from a major disability of simultaneously holding two conflicting jobs —heading a single agency while advising on governmentwide policies and programs. Skeptics doubted that the Director, who had to depend on the White House's good will for his own budget, could serve as a vigorous, independent counselor.

The Nixon Administration countered these arguements. Its spokesmen asserted that science and the scientific points of view were represented throughout government. Therefore, no need existed to bring these attitudes directly into the President's office. They contended that such advice already finds its way there on a daily basis. Moreover, OMB still puts together the federal budget, including its research and development components, still makes a special analysis of the total federal R&D effort, and receives the advice of pertinent agencies, including the NSF, in doing so. Administration spokesmen believe that the new organization, in reality, greatly simplified and made more effective communications links between those who work on science policy questions and those who implement policy decisions.

The Director insisted that he enjoyed the access to the President's top advisers. He expressed his confidence that he could see the President himself whenever he felt it was necessary. He also believed that the strengthening and increasing of his staff, especially the establishment of the Science and Technology Policy Office, helped him to perform his additional duties. Chairmanship of the FCST enabled him to exert influence in coordinating federal R&D programs.

The White House ultimately will have to decide if it wants to resurrect the old science advisory machinery in the near future. The effectiveness of the Director as an independent adviser remains to be seen. Yet, one can make two concluding comments. First, successful arrangements depend on a decisionmaker sympathetic toward them. There is nothing to be gained from forcing an advisory institution on a reluctant White House. Second, the President, as the final decisionmaker, should have the kind of staff support that he wants. He ultimately will have to take responsibility for both the quality of his decisions and the performance of the organizations that he designed to help him make these decisions.

The President also abolished the National Aeronautics and Space Council. This Council, under the leadership of the Vice President, had helped bridge a period of time when this country's space programs were being defined and organized. President Nixon felt that this country had effectively resolved its space and aeronautics objectives and had achieved the necessary coordination among agencies concerned. He, therefore, eliminated the Council without transferring its functions. This decision met no major opposition.

International Implications[8]

As indicated at the outset of this chapter, America's plunge into wholesale government support for science and technology stemmed in large measure from international considerations. Yet, this country's interest goes far beyond developing weapons to protect it from external threats. President Johnson, echoing the sentiments of three presidents before him, acknowledged that science and technology constituted potent tools for promoting international understanding. President Nixon shared this feeling, commenting that, "Increasingly, the peoples of the world are irrevocably linked in a complex web of global interdependence—and increasingly the strands of that web are

---

[8] See George C. McGhee, "International Scientific Cooperation: An American View," *Department of State Bulletin (March 7, 1966)* and U.S. Secretary of State, "Report on U.S. Foreign Policy in 1971: Section on Science," in *International Science Notes* (Washington, D.C.: U.S. Department of State, May 1972) No. 27, pp. 12-16.

woven by science and technology."[9] It is well recognized that basic scientific knowledge normally moves quickly across international borders. Individual scientists have long felt the need to communicate with foreign colleagues and have done so for centuries. On the other hand, advances in technology, engineering, and even medicine, have frequently been held back by lack of communications, indifference, varying social customs and beliefs, and occasionally, the desire for military or commercial advantage. It has been U.S. policy to facilitate even more the flow of scientific information and to accelerate the movement of technological information and devices throughout the world.

In meeting this objective, the United States supports a number of activities. In science, this support largely takes the form of government help for private activities, especially in assisting individual scientists and nongovernmental organizations. The latter institutions provide a common forum for the exchange of information among specialists. Probably the most important is the International Council of Scientific Unions (ICSU), the world's leading federation of nongovernmental scientific organizations, with a membership of over 60 nations. The best-known effort of the ICSU was the International Geophysical Year (IGY) in 1957-1958. This country proved a moving force within the ICSU in fashioning an Antarctic research program. In the early 1970s the ICSU, in cooperation with the World Meteorological Organization, planned the Global Atmospheric Research Program, designed to improve mankind's understanding of the circulation of the atmosphere. The first large-scale experiment in the field was scheduled to take place in the tropical Atlantic in 1974, and the first global experiment is scheduled to take place later in this decade.

Scientific cooperation has largely become the province of intergovernmental arrangements, both bilaterally and with multilateral international organizations. While most U.S. technological assistance has been administered through bilateral agreements, it has been showing increasing interest in multilateral organizations. There are about 60 such organizations—the United Nations Educational, Scientific and Cultural Organizations (UNESCO), and the Committee of Scientific Research of the Organization for Economic Cooperation and

---

[9] U.S. President, Richard M. Nixon, "The Importance of Our Investment in Science and Technology," House of Representatives Doc. 92193, March 16, 1972.

Development (OECD), being among the more active. UNESCO has a program for improving domestic science planning and establishing governmental institutions in member states. It also strives to improve the implementation of science planning, including the development of scientific communities and the initiation of priority research programs in natural resources and industry. The OECD has instituted a program of "scientific reviews" for each of its 21 members, thereby helping them fashion more effective science policies and institutions.

To aid in these efforts, the Department of State, responsible for all official foreign relations of the United States, created the Office of International Scientific and Technological Affairs and assigned scientific attachés to 18 embassies around the world, to the OECD mission in Paris and to the UNESCO mission in New York. Responding to a congressional directive, in 1974 the Department of State established a Bureau of Oceans and International Environment and Scientific Affairs. This new bureau expands the responsibilities of the old Office of International Scientific and Technological Affairs to include such activities as oceans, fish and wildlife, and population, previously handled by other organizations of the State Department.

The United States has sponsored international cooperation in a variety of scientific fields. In the field of civilian atomic energy, since 1953 this country has shared its nuclear science with other countries. It has done so both as a moral obligation and to share the expanding world market in nuclear goods and services. Peaceful applications can be broadly divided into three categories—nuclear reactors for energy, radioactive isotopes, and nuclear explosions.

In 1971 this country had 34 agreements for cooperation, including those with the European Atomic Energy Community, provided scores of research and power reactors using enriched uranium, and became the chief supplier of the nuclear fuels to power them. In 1971, in a major policy change, the United States decided that under certain conditions it would be willing to make its gaseous diffusion uranium enrichment technology available on a selective basis to other nations. This country has explored with other governments the possibility of utilizing American gaseous diffusion technology in a multinational plant for peaceful purposes. It insists that the project would be under international safeguards with adequate security provisions to protect the sensitive technology and expects reasonable compensation for the use of its technology.

For many years the United States has supported the efforts of the International Atomic Energy Agency to apply safeguards to prevent the diversion of nuclear materials from peaceful to weapons programs. Thus it has aided the IAEA to develop the safeguard system it requires to carry out its responsibilities under the Non-Proliferation Treaty.

NASA also has invited foreign participation in the U.S. space program. Some countries have agreed to joint satellite design, development, and launch projects. Any foreign scientist may suggest experiments for inclusion in satellites developed by NASA. With U.S. assistance, 14 European satellites have already been launched and are in orbit. More are planned for the future. The United States has included foreign participation in its earth resources satellite program. These satellites are designed to observe oil and mineral deposits, agricultural crop yields, plant diseases, forest management, and the like. Scientists from 22 countries have contributed proposals to this NASA program, and the data acquired will be made available to all nations requesting them.

The United States has promoted cooperative efforts with the Soviet Union. By 1971, NASA and the Soviet Academy of Science exchanged samples of lunar soil and established joint working groups in areas such as meteorology, biology, medicine, the study of the natural environment, and space. In 1972, President Nixon and Soviet leaders in Moscow agreed to promote cooperative efforts in the fields of environmental science, biomedicine, and space. They specifically agreed on a great adventure in the cooperative exploration of space with a joint orbital mission, including a docking maneuver, of an Apollo and a Soviet spacecraft by 1975. The two nations set up a U.S.-U.S.S.R. joint commission to review periodically progress achieved by this science exchange program.

President Nixon has used science as one of his major tools for improving relations with the People's Republic of China. During his visit to Peking in 1972, he and Communist China officials agreed to facilitate increased people-to-people contacts between scientists and engineers (among several professions) of the two countries.

The United States carries on much of its international science programs through bilateral agreements. These agreements normally involve exchanges of information, consultation among scientists, and practical cooperation in the laboratories of the nations involved. Operating under a bilateral agreement. American and French scientists are working in each other's laboratories as well as cooperating in carrying out studies of making the French city of La Vaudreuil pollution free. They also are collaborating in developing an improved heroin detecting device, an important investigation considering the major drug problem now confronting both nations.

Instruments of Government Support

Mushrooming scientific and technological needs have prompted government to augment the activities of its own laboratories by contracting with private organizations. In certain cases, and for a variety of reasons, private institutions such as industry, foundations, universities, and nonprofit organizations, are preferred to perform scientific work rather than government laboratories. These organizations help the United States extend its resources beyond its own facilities. The chief instruments for sponsoring this work include contracts, grants, patents, educational support, and dissemination of scientific information.

The contract system represents one of the oldest methods for government procurement of goods and services from private organizations. It is, in essence, an improvised form of federalism that incorporates both governmental direction and regulation and a great deal of freedom for local private enterprises. An administrative contract allows a private corporation considerable independence in planning and implementing programs, and a degree of flexibility which is often not maintained in the more rigidly administered federal agencies. In procuring R&D, in contrast to manufactured items, uncertainty greatly influences contract types. Consequently, a variety of contracts has been developed specifically for providing a legal and administrative working relationship between government and private R&D organizations.

In addition to the more traditional R&D contractors—industrial firms, universities, and research institutes—some new types of contractors arose after World War II. The nonprofit organization emerged to provide the government with the services of scientists who, for a variety of reasons, preferred to work outside federal agencies. The RAND Corporation, for instance, performs such a function for the Air Force and other agencies. In addition to research in basic and applied sciences, these organizations often provide policy research and managerial advice.

Another postwar development is contracting with private institutions to manage special research centers. The Sandia Corporation, specializing in research on atomic weapons, is managed by the Western Electric Company under contract to the federal government. The University of California manages the AEC-owned Livermore and Los Alamos Laboratories.

Federal grants differ from contracts in that they usually support fundamental research in a given field rather than a specific research project directed toward solving a directed problem. While the system of providing grants is undergoing major changes, many federal agencies, most notably the National Science Foundation, support basic research through this mechanism.

To encourage inventions, the federal government, since 1790, has operated a patent system whereby inventors receive exclusive right to make, use, and sell their inventions for a number of years. The Patent Office has created a central depository of patented and unpatented models and specimens of manufacture. It has constantly employed technically qualified examiners to test each invention. By establishing a system of patents which rewards men and organizations for their findings, the government encourages scientific pursuits.

Another important instrument for promoting science and technology is educational support. The federal-university relationship also involves contracts to support federal programs, and grants to support research or to improve the quality of scientific education.

Supporting the effective use of information media has become a major task of goverment in promoting science and technology. The government has organized several technical communication systems, some of which include research in nongovernmental organizations and foreign countries. In this manner the government not only helps researchers engaged in its own programs, but also becomes familiar with essential facts concerning the programs of professional and technical societies, private industries, and universities. Improved communication also aids the government in preventing unnecessary overlapping of research activities. Information specialists are attempting to unify the communication systems in order to make them more effective.

Summary

From the founding of the Republic to the present day, government has shown a growing concern for science and technology. The reasons for this trend are easy to find. Science and technology increasingly have become crucial in assisting this country in maintaining its security, promoting its economy, and protecting the health and safety of its people. As never before in history, the status of science and technology has become an important hallmark of a nation's greatness; and the United States clearly has perceived and acted on the basis of this fact. In the process, the federal government has displaced the university, industry, and the private foundation as chief patron and has fashioned a host of institutions to administer its vastly increased commitment to scientific and technological excellence.

6 Selected Problems and Issues

Ralph Sanders

Before exploring the problems of science policy, a brief semantic note is required. In this chapter, discussions will include within the term "science" both science and technology except when the distinction between the two is important to an understanding of the subject. The purpose for this semantic shorthand is to avoid a constant reiteration of both terms. The use of this shorthand device has already become commonplace with the public. For example, the man-on-the-moon project frequently was labeled a "scientific" enterprise although the major effort was technological. Its use here is not to contradict the distinctions between science and technology pointed out in Chapter 1, but to take advantage of economies of language.

The preceding history of the relationship between government and science documents a truly remarkable record of achievement. Increasingly, the United States has promoted science to help it solve its problems. Granted, there were failures along the way, but successes proved particularly outstanding and seem to have had a lasting effect.

Yet, major problems remain, generating issues vitally affecting American society. Indeed, the application of science as demanded by our society has brought unto the American people some major problems. It is impossible to explore here every problem. Consequently, this chapter will be limited to a select few that appear to have major impact.

Decisionmaking and Substantive Issues

One can categorize two general types of problems associated with the social dimensions of science and technology. The first relates to the decisionmaking task itself and concerns the institutions and techniques used in formulating and implementing policies. The second concerns substantive issues that these policies must address. Let us first examine some major problems and issues associated with the decisionmaking process.

Science Advisers

What should be the role of the science adviser to government? These advisers, both individually and in groups, hold positions throughout the decisionmaking structure, from the highest to the lowest echelons. While the federal government has hired many scientists as permanent employees, many others serve part time while retaining their positions in the private sector. In general, these advisers have served their government admirably. Yet, government's use of science advisers has raised some perplexing problems.

Choosing advisers to serve government poses no easy task. For the most part, the federal government uses an informal selection process, based largely on the personal acquaintances of those scientists already working for or advising government. Consequently, care must be taken to avoid self-perpetuating particular views and prejudices. It is all too easy for a scientist of one persuasion to recommend scientists of similar beliefs to advisory posts. As an example, for many years critics charged that the President's Science Advisory Committee was largely composed of physical scientists from a few of the nation's leading universities. Over a period of time the membership of the PSAC changed to include men from the biological sciences, from the social sciences, from industry and other nonacademic institutions, and from a larger number of universities.

The part-time science adviser faces a dilemma regarding his responsibilities. Should he represent the scientific community to the government or look out solely for the government's interest? Most observers agree that advisers should not act as official spokesmen for science, but as loyal members of governmental organs. Yet, a part-time adviser cannot abruptly shift his allegiance when he takes on a counseling job for government, nor in all cases can he distinguish sharply community from government interests. Happily, the interests of government and his profession need not conflict. Nonetheless, sorting out one's fealty remains a continuing problem for the part-time science adviser.

Scientists also encounter problems in communicating with policymakers. This difficulty inevitably results when men have to explain complex, and to some degree esoteric, issues to laymen. Consider the difficulty of explaining the nature of a nuclear accelerator to a politician. Probably the best that the scientist can do is illuminate the very general principles involved while pointing out the political, economic, cultural, or social impact. By and large, both executive branch administrators and congressmen have come to understand enough of natural phenomena and scientific disciplines to judge the social and political worth of research.

Other problems arise relating to the role of the scientific adviser. One of these concerns the question of prestige versus technical competence. Sometimes well-known scientists lend their prestige to support certain scientific proposals. There inherently is nothing wrong or sinister about this practice. A well-qualified scientist can pass judgment on other than his own immediate project. If he truly believes in its merits he performs a valuable service calling this project to the attention of decisionmakers. On the other hand, if he backs a project that lies well outside the fields in which he specializes, one could question his credentials to evaluate the proposals. The political leader faces a particularly acute dilemma when two highly respected scientists take opposite sides on the same issue. In such cases, ultimately he has to reply on his own common sense to arrive at a sound decision.

Resource Allocation

No issue involving science commands more attention than that of allocating funds to support competing R&D programs. Not surprisingly, more often than not, success comes to those lines of investigation that received ample funds. Each scientist, research institution, and government agency knows this fact full well and, thus, competes fiercely for as large a share of the resources as he possibly can get. Responsible decisionmakers, charged with managing R&D enterprises, likewise know that the future of their organizations as well as that of the nation may depend, in part, on the resource allocation decisions that they make.

No formula exists for ensuring an optimal allocation of resources among fields (basic, applied, or engineering development), disciplines (chemistry, biology, physics), or projects (space shuttle, earthquake investigations, lasers). There can be none, because the goals of society involve differing subjective values, making it difficult for men to arrive at consensus. Whether one decides that mental health merits more support than nuclear energy rests on more than purely objective considerations.

We simply lack commonly accepted criteria for making tradeoffs between scientific pursuits. For example, if in our health research we stress achieving the economic objective of reducing time lost by the work force, we might be induced to allocate large sums to research designed to cope with the common cold. But, a congressman who has had a family member die of cancer would look askance at an HEW request for money to study sniffles.

Although the federal government has set up no standardized national priorities for choosing those science enterprises most relevant to the nation's needs, various individuals have proposed schemes for this purpose. The scheme advanced by Michael Reagan is representative of such ideas and sheds light upon the problem, even if it offers no proven solution.[1]

Reagan lists four priorities. First priority should be given to those objectives which are defined politically as most urgent and to which scientific research can most clearly make a contribution. Science-related educational needs, from elementary schools to graduate education-laboratory research, should receive second priority. Third comes support for a healthy general base of undirected small-scale research. This activity represents the investment capital upon which rests utilitarian technology and the application of science to areas of social needs, and, according to Reagan, it does not receive a very high priority. Lastly, come fields having very high equipment costs and devoted to furthering basic knowledge. Thus, "big" science requires a few, but expensive facilities and sometimes its payoff lies in increasing our understanding of the universe rather than in meeting societal needs.

Neither this scheme nor any other probably can become a universally accepted mechanism for making allocation decisions. In our pluralistic governmental system, characterized by conflicts of interest, no such scheme probably would receive the support of all. In addition, we still have not developed the techniques to make such decisions in a purely rational way, excluding subjective values from consideration. Some people have suggested using operations research or systems

---

[1] Michael D. Reagan, "R&D: Suggestions for an Allocation Framework," *Public Administration Review*, Vol. XXVII. No. 2 (June 1967), pp. 104-111.

analysis techniques for such purposes. The fact is that such analytical techniques have proven useful in aiding decisionmakers to choose from among technological options. Cost effectiveness studies do offer opportunities for making a more rational selection as among competing technologies designed to achieve a common purpose. However, these techniques are designed more to shed light on how to utilize resources than on arriving at the goals themselves. This fact limits their application in setting technological priorities, and makes them almost inoperative in choosing from among basic scientific investigations (which by definition are not goal directed). Yet, it is the continuing controversy over goals that most affects resource allocation.

Consequently, we shall continue to face resource allocation problems as well as the pulling and hauling of the bureaucratic and political systems in which these decisions are made. The most likely approach to making better decisions is to improve our budgetmaking systems, including assuring that all competing interests have an opportunity to make their cases known.

Coordination of R&D Programs

Since the birth of the Republic, there has been a conflict between centralizing and decentralizing tendencies in government decision-making. As science and technology became significant items in government operations, this conflict began to affect them. As indicated in Chapter 5, the United States has not consolidated the management of its science and technology into a single organization such as a Department of Science and Technology. Rather, each federal agency dedicates its R&D efforts to further its particular mission. Each of these missions does not demand a unique set of sciences or technologies, clear and distinct from all others. Often agencies pursue similar lines of investigation. For example, at least eight agencies have a major interest in conducting research in the field of ocean sciences, including the Departments of Defense, HEW, Interior, and the National Science Foundation. Likewise many agencies sponsor the much needed energy R&D, including the Departments of Interior, Defense, Commerce, and the Atomic Energy Commission.

Not all duplication is regarded as harmful. Some kinds are deliberate and warranted, representing different attacks on the same problem. Others flow naturally from the nature of the R&D process; still others may prove undesirable, but unavoidable. Helpful duplicate research occurs frequently in basic research (where costs per project are small). Yet, since duplication eats up scarce resources, government has an obligation to keep it to a minimum unless some overriding reasons exist. Closer topside management aims to prevent duplication resulting from two or more agencies simultaneously conducting identical research. Just as important, it seeks to make the efforts of participating agencies complementary and mutually reenforcing.

However, coordinating simply for the sake of coordination contributes inefficiency and could damage worthwhile R&D enterprises. When agency officials find it necessary to spend an inordinate amount of time in liaison with other agencies, coordination could prove counterproductive. Nonetheless, the need to coordinate remains as long as mission agencies undertake various aspects of similar research tasks.

Government officials face many possible alternatives in achieving coordination. They might desire, for instance, to designate one agency as the funding organization, giving it responsibility for directing the entire effort. This option raises the question whether any agency should direct an effort if its contributions represent a minor portion of the overall R&D resources needed or if the effort constitutes only a small portion of that agency's total R&D commitment. When a field of activity overlaps several agencies, none having it as a primary mission, perhaps an independent committee could prove the superior solution. Another alternative might be to have fewer agencies or only one engage in research in this particular field. The latter option presents the question whether the interests of excluded agencies would be observed, neglected, or adequately emphasized.

Chapter 5 pointed out that the Federal Council for Science and Technology (FCST) came into being with the purpose of helping to coordinate federal science programs. In many respects the FCST has promoted cooperative and integrated management. It also has facilitated communications among participating agencies in what has to be a pluralistic system. Yet, FCST has advisory rather than enforcement powers. For example, while President Eisenhower's order setting up the FCST granted that body advisory functions relating to the use of capital, the Council has never attempted to prepare budget recommendations. Moreover, agencies sometimes have tried to use the Council to legitimatize their own views, programs, and budgets by seeking its stamp of approval.

The problem of coordination should be kept in perspective. The informal mechanisms for coordination, especially the innumerable daily personal contacts among managers and researchers of various agencies, still have a profound influence. And despite the tendency of agencies to work for their own self-interest, they generally still are willing to work together for the common good. Still enough differences remain to require careful consideration of problems of coordination. In his new role as science adviser, the Director of the NSF will have to pay considerable attention to this problem.

Substantive Problems

There are two major types of substantive problems—generic and specific. The generic problems affect society in a broad way; their consequences are far reaching. It should be pointed out that most generic problems result more from technology than from fundamental research. Let us first look at some major generic problems and then turn our attention to specific ones.

The Wisdom of Economic and Technological Growth

Three closely related questions recently have tended to dominate the debate about the value of science and technology to mankind: whether continued economic and technological growth benefits or harms society; whether the depersonalization that technology allegedly triggers has inflicted too high a human cost; and whether the rise of antitechnology attitudes omens a general social backlash that could impede further progress?

The first question has received enormous publicity during the past few years. It is a well known fact that the more scientifically and technologically advanced a society, the more effective is its economic performance. R&D make significant contributions to economic growth by stimulating improved production methods, new products, and sometimes even new industries. For example, the computer, synthetic fibers, electronics, and aerospace industries did not exist 70 years ago; today they constitute major pillars supporting the economic structure of this nation.

Above all, the application of science and technology has enabled men to increase their productivity, that is, their ability to produce more goods and services with the same or less effort. In other words, it has permitted the output per man-hour to go up. The result has been higher standards of living with less inflation. However, within recent years the annual growth rate of productivity has declined in the United States, from 3.4 percent between 1940 and 1966 to 1.6 percent between 1966 and 1970, with projections of an even greater slowdown in the future. If one values economic growth, he could conclude that the nation needs an accelerated infusion of research and development to turn this unwelcome trend around.

It is very difficult to ascertain with any precision the contributions that science and technology make to economic progress. We really don't know how to compare their contributions to those of other elements like land, labor, capital investment, and management. Yet, few would contest the fact that they made valuable contributions. We observe that countries with a high per capita gross national product (GNP) have a substantially higher research ratio than those with a low per capita GNP. Advanced industrial countries typically spend more than 1 percent of their GNP on R&D while developing countries spend less than 0.25 percent. Thus, one can conclude that a positive correlation exists between investment in research and development and economic advancement.

Yet, increasingly, many men have come to question the need for economic growth, and as a corollary, for the technological expansion that helps fuel it. As the Organization for Economic Cooperation and Development recently noted, "Faith in economic growth has been replaced by a feeling of unease in the face of the prospects opened up by it and has been shown to be insufficient in itself to respond to the aspirations of mankind for a better way of life. What is more, far from bringing only benefits, it is accompanied by more or less traumatic effects, less and less acceptable to large sectors of public opinion."[2]

In 1972, a computerized study by some analysts at the Massachusetts Institute of Technology concluded that if mankind continues the current exponential growth rates of population and capital, collapse will

---

[2] Organization for Economic Cooperation and Development, *Science Growth and Society: A New Perspective* (Paris: OECD, 1971), p. 26.

occur in about 50 years.[3] These analysts developed and studied a world model and examined the interacting consequences of continued growth of population, food production, industrialization, pollution, and the consumption of nonrenewable natural resources.

Perhaps no adverse consequence of economic growth has commanded more attention than pollution, and certainly the MIT study gave it considerable stress. While some conservationists long have warned against the buildup of harmful wastes, it was Rachel Carson's book, *Silent Spring*, in 1962, that initiated massive public outcry. She marshaled evidence to show that DDT was causing severe ecological damage. Barry Commoner, presently sounding the ecological alarm, has aptly summed up the environmentalists' case. "This pattern of economic growth is the major reason for the environmental crisis. A good deal of the mystery and confusion about the sudden emergence of the environmental crisis can be removed by pinpointing, pollutant by pollutant, how the postwar technological transformation of the United States economy has produced not only the much-heralded 126 percent rise in GNP, but also, at a rate about ten times faster than the growth of GNP, the rising levels of environmental pollution."[4]

The MIT scholars say that the only way to prevent disaster is to halt economic growth and to aim for economic stability (or equilibrium). Such a course of action must affect significantly the handmaiden of economic growth, technological progress. While these analysts do not argue for an indiscriminate stoppage of technological advances, they do strongly suggest that men reduce their support of technological efforts that tend to increase the size of the world's capital plant. They contend that despite the views of optimists, technological therapies would come too late to avert demographic or environmental catastrophe, and at best could only delay rather than avoid crisis. Therefore, they argue that in addition to stopping population growth, man must hold down his overall investment in scientific and technological innovation, and direct his more limited research and development activities to those pursuits that would enhance the workings of

---

[3] Donella H. Meadows, Dennis L. Meadows, Jorgen Randers, and William W. Behrens, *The Limits to Growth: A Report for the Club of Rome's Project on the Predicament of Mankind* (New York: Universe Books, 1972).

[4] Barry Commoner, *The Closing Circle* (New York, Knopf, 1971), p. 146.

a steady (nongrowth) society. Man can live indefinitely on this earth only if he imposes limits on himself and his production and consumption of material things.

Other observers take a less pessimistic view, preferring to believe that there are ways for society to promote the beneficial effects and to mute the harmful consequences of the application of science and technology. They argue that halting our investment in R&D enterprises does not seem a responsible solution. A more sensible remedy, according to this school of thought, would be to control science and technology as well as their social consequences, at least to a tolerable degree. While technology can unleash unforeseen, unwelcomed results, man has the capacity to employ this resource as he wants, to advance or harm society. He can so direct its use and prompt the kind of economic growth that would enhance rather than diminish the meaning of his existence.

In other words, man can find the right uses of science and technology. He is not helplessly craven before an invincible, malevolent technological force. President Nixon seemed to recognize this proposition as a truth when he observed, "the impact of new technology can do much to enrich the quality of our lives." [5] This sentiment is not politically partisan. Senator Edward Kennedy likewise noted that, "Like it or not, science and technology have become central to our civilization, to our economic strength, to the preservation of our environment, and to the quality of our lives." [6]

Those expressing more optimistic (and in their view, realistic) views argue that deliberately calling a halt to economic growth offers poor counsel for a number of what seem to them to be convincing reasons.

First, they would contend that there are too many people in both advanced and developing societies who want to move out of poverty, and they must have advanced technology to help them do so. These would include technologies that help to build capital plant and products. Some idealists have suggested that developed nations export part of

---

[5] U.S. President, Richard M. Nixon, *The Importance of Our Investment in Science and Technology,* House of Representatives Doc. 92193, March 16, 1971.

[6] U.S. Congress, Senate, Subcommittee on Computer Services, Committee on Rules and Administration, *Office of Technology Assessment for the Congress: Hearings* (Washington, D.C.: U.S. Government Printing Office, 1972), p. 38.

their accumulated wealth to underprivileged countries while holding down their own growth. It is open to question whether advanced societies will display such selflessness and generosity, if for no other reason than that they, too, have their own pockets of poverty to eradicate.

Second, it is against human nature to forgo the hardware that makes life more interesting and less arduous. It is very difficult, but not necessarily impossible, to exact economic penalties that force people to give up or forgo technological improvements. Certainly poor people in the inner cities and rural areas want automobiles, television sets, and other fruits of technology. Peasants chained to subsistence farming likewise yearn for more and better material possessions. Most likely the countries of Latin America, Asia, and Africa will not abandon efforts to get all the technology that they can. Since it is mostly Westerners who call for halting economic growth, many people in less developed nations view this exhortation as another form of imperialism.

Third, critics of the MIT study probably would acknowledge that in a sense the MIT analysts are right in asserting that the resources of the world are finite. The recent worldwide energy crisis could be considered as concrete evidence of their correctness. Those who claim that technology can conquer all most likely are on shaky ground. Yet, these critics assert that we can expect more from technology than the MIT scholars suggest. They question, for example, the MIT study's apparent assumption that what we consider resources today will remain the only resources for the future or that we cannot get more out of these resources later than we do now, and at a rate that will avert disaster.

In the past, by and large, technology has met the challenge by enlarging our stock of useful resources. The MIT study acknowledges this fact. When rubber became scarce, synthetic rubber took its place. When it was thought that famine would sweep the less developed world, the "green revolution" helped to avert massive catastrophes. (Of course, it has not helped eradicate all hunger and deprivation.) Overall, the performance of science and technology has torpedoed the more frightful Malthusian predictions.

Can they continue to do so? The more optimistic contend that with enlightened social policies, they probably can. For example, these observers would tend to believe that the energy research and development measures discussed in Chapter 4 could very well produce the answer to the long-term energy shortage faced by the United States, and perhaps the world. To them it is wrong to assume automatically that our usable resources are forever fixed and will run out. Moreover, they tend to think that technology will provide commercially feasible energy sources in time to prevent major economic and social dislocations.

Fourth, believers in a more prominent role for investment in science and technology place greater emphasis than do the MIT scholars on the potential contributions of science and technology in facilitating beneficial transformations in the compositions of economic growth. As the MIT analysts have acknowledged, they could lead to products and entire industries devoted to cleaning up the environment and keeping it that way. Improved waste disposal plants, pollution abatement devices, improved health care instruments, mass transportation systems, more abundant and more nutritious food are some of the needs that will depend, in large part, on what society does with its science and technology. These technological advances could add to both economic growth and the capitalization of society.

In sum, pessimists have little faith in man's ability to control the adverse consequences of science and technology and fear the impact of economic growth. On the other hand, the more optimistic feel that over the long run there is little likelihood that men will deliberately slow down either economic or technological growth, but rather will seek to channel them in ways that accord with society's sense of well being. One can only hope that social control will meet the challenge.

Depersonalization

The argument that technology depersonalizes human beings has gained many adherents in recent years. Marx contended that the factory system, with its repetitive and standardized motions, robbed man of his humanity. Today, critics charge that mass production, automation, and the computer have relegated man to an unimportant cog in a vast machine. Lewis Mumford, a highly respected scholar, asserts that within the past four centuries our traditional system of life "has been replaced by a system that gave primacy to the machine, with its repetitive motions, its depersonalized processes, its abstract

quantitative goals." [7]   He went on to assert that "man himself is losing
hold on any personal life that can be called his own." [8]

We see the same sentiments expressed by youthful protesters who
rebel against treating people like punch cards. That is the message
in the signs they carry reading, "Do not fold, spindle, or mutilate [us]."
If man has offered his soul to a technological devil, society indeed faces
a bleak future. Yet, it is important to point out that we are dealing
here more with images than with fact based on concrete evidence.

Certainly some technologies, like mass production, have tended to
reduce man's individuality. The assault on privacy by the computer
has troubled many men and prompted President Nixon to propose
study and action to avoid this unwholesome development. Yet, equally
convincing arguments contest the conclusion that technology inevitably
must rob man of his personality. Pre-high technology societies did not
afford man maximum opportunities to develop unique character. The
monotony and danger of subsistence living offer the impoverished
peasant little chance to fashion his own personality. There is no
evidence that the Pilgrim fathers, unencumbered by advanced technology, exhibited greater individuality than even the workforce of
General Motors. Most probably, the Pilgrims demanded greater con-
formity. In fact differences in dress, belief, and lifestyle today vary
more than they ever have (even after the enormous technological
onslaught of the past three decades).

In many ways science and technology, it could be argued, have given
men more opportunities to fashion their own personalities. They have
helped make possible a more educated population. Increased knowledge multiplies the probabilities for greater diversity. Moreover, technology has helped bring about a leisure revolution. Men now follow
a wide variety of avocations that tend to make some people different
from others. Today, a man working in an automobile assembly plant
can visit Europe, experiencing sights and cultures that even 40 years
ago were available almost exclusively to the nation's rich and elite.
One could argue that while certain technologies do tend to deperson-

---

[7] Lewis Mumford, *Myth of the Machine: The Pentagon of Power* (New York:
Harcourt, Brace, Javanovich, 1970), p. 164.

[8] Mumford, p. 287.

alize in some situations, we cannot conclude that as a rule technology leads to a loss of human qualities. Moreover, we can work, according to this view, to remedy the depersonalizing tendencies of technology.

Antitechnology Attitudes

Of late, a rising countercuture has engulfed many people, especially the young. Concerned about pollution, fearful of nuclear holocaust, frustrated by urban congestion and poverty, alarmed at the depletion of resources, militant critics have gone on antiscience and antitechnology crusades. They not only have questioned the uses to which these resources have been put, but the wisdom of expending efforts in these directions.

There is no better illustration of this feeling than the commonly heard criticism: Why should we have spent money putting a man on the moon when millions of people live in poverty? Often there is a call for more welfare, better health care and many other services that involve little in the way of science and technology. There is no intent here to argue the merits of these alternative uses of our resources. The point is that some people want to divert many resources devoted to our technological investment to these other uses, partially in the belief that technology produces pernicious effects.

Of interest, the revolt against technology has not been spearheaded by workingmen whom the machine would displace first. We are not witnessing another early 19th-century Luddite uprising when bands of displaced craftsmen sacked English factories, destroying machinery. While it is true that workingmen and their unions have resisted certain technologies, intellectuals, especially on American campuses, have fired the most intense salvos against technology. Scholars and students have proved the vocal architects of the antitechnology crusade.

An opposing view holds that most likely the size and influence of the counterculture have been exaggerated. Moreover, it seems to have passed its peak. No doubt, more people than ever before now question the wisdom of technological progress. Most people want to prevent the adverse consequences of certain technologies—belching power plants, for instance. But, according to this opinion, there is no widespread movement to banish technology. Even the most severe critics of air pollution often go to and from their protest meetings in automobiles.

Technology Assessment

The real issue is: How can society prevent science and technology from triggering unforeseen, adverse social consequences? In addressing this question a subsidiary question arises: Can man consciously plan his science and technology so as to ensure that he doesn't reap a bitter harvest? Thoughtful men have pondered this question long and hard. One result of this concern has been society's demand for environmental impact studies for proposed new technological projects. Environmental impact studies constitute one type of systematic analysis designed to control technology, usually labeled "technology assessment."

Technology assessment is the examination of the potential impact of technology on society in order to minimize its probable, harmful side effects. Analysts attempt to delineate options and costs, encompassing economic as well as environmental and social considerations. In 1972, Congress, long concerned about controlling techology, set up a congressional Office of Technology Assessment.

Most literature related to technology assessment concentrates on discerning undesirable effects, the difficulty of overcoming them, and how to organize the technology assessment enterprise. What is lacking are the analytical tools to examine these problems systematically. Given the uncertainties, men find it difficult enough to forecast future technology. To estimate the associated social consequences poses an even harder task. Moreover, observers have pointed out the difficulty of ensuring objectivity and of making sure that analysts take all important factors and views into consideration. Some of its proponents acknowledge that technology assessment remains an imperfect tool and guarantees no panacea. In fact, some scientists feel that technology assessment, if narrowly construed, could warp the creative development of American science and technology. To them, technology assessment could easily turn into technology harassment.

Despite these deficiencies and misgivings, some kind of technology assessment seems here to stay. Many people agree that if it contributes nothing more than focusing attention on the problems of preventing harmful consequences, society will benefit. This increased attention, even with crude analytical tools, at least offers us an opportunity to use our brains in an attempt to control technology and its uses.

Basic versus Applied Research

We now turn to some specific problems and issues that confront our policymakers.

Some critics lament the alleged preference of governments for funding basic over applied research; others take an opposite view, expressing fears that basic research is being starved. Those that would like to see applied research emphasized more contend that social benefit, the objective of applied efforts, should take priority over simply acquiring new knowledge, the goal of basic research. Moreover, they argue that rarely does a solution to social problems come directly from fundamental investigations. As one critic observed, "As a rule, problems are solved by engineers, doctors or applied psychologists, not by physicists, biologists or experimental psychologists." [9] He goes on to point out that the most important advance in drug abuse therapy was methadone which was introduced by two doctors who found it practical and inexpensive.

Our national leaders, likewise, have recognized the need to harness science to help solve major problems, especially in the domestic field. President Lyndon Johnson on several occasions urged U.S. scientists to concentrate on practical problems. In his State of the Union Message in 1972 President Nixon commented: "I have felt . . . that we should . . . be doing more to apply our scientific and technological genius directly to domestic opportunities. . . . We will seek to set clear and intelligent targets for research and development, so that our resources can be focused on projects where an extra effort is most likely to produce a breakthrough and where the breakthrough is most likely to make a difference in our lives." [10]

In 1968, Congress enacted legislation permitting the National Science Foundation to strengthen its ability to support applied and social sciences. In response to both congressional and Presidential desires, the NSF inaugurated a program of Research Applied to National Needs (RANN).[11] This highly selective program pursues additional scientific

---

[9] Amitai Etzioni, "Redirecting Research Dollars," *The Washington Post,* June 11, 1972, p. B-3.

[10] U.S. President, Richard M. Nixon, "State of the Union Message," *Vital Speeches of the Day,* Vol. 38 (February 1, 1972), 226-229.

[11] See also Chapter 5, "Federal R&D Institutions."

knowledge in a liimted number of areas that hold promise of technological, social, or economic payoff. As part of RANN, scientists are
evaluating future energy needs and alternative methods of supplying
this energy. The energy crisis of 1973-1974 prompted increased emphasis in this area. RANN scientists also are looking into developing
the use of enzymes for industrial purposes to assure higher efficiency
and no pollution.

Yet, critics remain dissatisfied with this trend and call for an even
greater shift in emphasis from basic to applied research, implying that
a problem of resource allocation still exists.

Defenders of basic science see danger in orienting our total research
effort exclusively toward solving social problems. First of all, they
contend that the level of basic research is not excessive. Overall, as
noted in Chapter 2, basic research accounts for only some 15 percent
of total R&D expenditures. The federal government spends a similar
percentage of its R&D budget on fundamental investigations. Consequently, there is relatively little margin for significant reductions. Yet,
in absolute numbers the United States spends over $4 billion for basic
research, with federal contributions totaling some 63 percent. Thus,
fairly large sums of money could be diverted from basic to applied
research.

Yet, no one can say with complete confidence that monies could be
profitably reduced from basic research or how great a reduction it can
take without damaging the nation's overall scientific enterprise. Many
supporters of basic research argue that in addition to satisfying man's
intellectual curiosity, basic research provides the source of fundamental knowledge that applied researchers will tap for years to come.
President Nixon recognized this fact in stating, "we must continue to
give an important place to basic research and exploratory experiments
which provide the new ideas on which our edifice of technological
accomplishment rests." [12]

Furthermore, as long as the federal government continues to fund
university education and training for scientists, it simultaneously will
continue to support basic research because most university research
is basic. The question of basic versus applied research will continue

---

[12] U.S. President, Richard M. Nixon, "The Importance of Our Investment in
Science and Technology." House of Representatives Doc. 92193, March 16, 1971.

to cause controversy, but most probably the effort devoted to the former will neither increase nor decrease appreciably in the foreseeable future.

Higher Education

Since World War II the federal government has become a major supporter of science education and research conducted by the nation's universities. Large-scale federal financing has literally changed the nature of universities, affecting their capacity to teach, changing their financial status, establishing new political relationships, and influencing the organization of research. American universities have long used research as a major tool in their science instruction. An important method for training graduate students is to allow them to participate in actual research. Consequently, it is argued that a major reason for federal support of research is to aid in producing more and better-trained scientists. In addition, universities have proved uniquely qualified to conduct certain kinds of research that government might need.

The close relationship that grew up between the federal government and the universities produced many benefits, but also triggered certain problems. In allocating resources, for instance, federal science administrators constantly face the difficult task of deciding whether to fund projects that stress research results or that emphasize education value. The most deserving research in terms of discovering new knowledge may not be the best way to train scientists. Furthermore, the usual problems remain—deciding which project has the greatest promise of contributing to our understanding of nature or of producing social benefit.

There is no intention here to explore all problems associated with federal support for university research. Instead, this discussion will focus on three issues: (1) the decline in federal support for university research, (2) the wisdom of shifting emphasis from project to institutional support, and (3) the validity of defense-sponsored research on campuses.

Universities and colleges rely upon outside sources for support for more than three-fourths of the total funds they expend for R&D. The federal government became a major sponsor, and educational institutions came to rely on its largesse. From the end of World War II a steady rise took place in the federal share of total R&D spending among institutions of higher learning. Since 1968, however, the federal portion levelled off at about three-fifths of total academic R&D. While federal funds continued to increase, their annual growth rate fell to 3 percent between 1968 and 1970 compared to 14 percent between 1964 and 1968. In constant dollars, federal funds actually declined 2 percent since 1968 because, faced with sizable inflation, universities could buy less and less research for each federal dollar. Nonfederal sources of support did not compensate in any major way for this slowdown in federal contributions.

As a result, within recent years science departments found themselves operating on austere budgets. Fellowships devoted to educating scientists began to shrink. University scientists and officials decried this drop in funds, charging it would damage the nation's scientific and technological competence in the years ahead. The federal government responded that money was tight, that other sectors of national life faced similar financial difficulties, and that other needs also had to be met. Although for 1975 an increase in federal support across the board is contemplated, universities still are having to learn to make the most effective use of the resources that they do receive as well as to augment them with nonfederal contributions.

Institutional Support

About 80 percent of all federally sponsored research is conducted on campuses of 164 public and private universities and colleges. The federal government has supported such research chiefly through funding specific projects either by means of grants or contracts. Thus, individual scientists who propose these projects and carry them out, find themselves competing fiercely for the federal dollar. Financing university research as a whole (rather than individual scientist members of their faculties) has received less attention.

Critics charge that heavy reliance on project funding leaves much to be desired. They believe that project grants do not provide the relatively unrestricted funds that an academic institution needs to balance its programs, upgrade its offerings, strengthen its faculty, revise curricula, and establish supporting services. Targeted project grants, which may serve to meet currently pressing needs, may also serve to divert the energies of faculty members and the institution's physical facilities away from basic research.

As a result, many university administrators have advocated supplementing project support with heavier stress on institutional grants. Under this concept, universities receive a block of funds which they can allocate as they see fit. Such support, according to its promoters, would contribute a stable, long-range funding for research and instructional programs. The university would no longer be so heavily dependent on the salesmanship of its individual scientists in securing federal monies. With institutional grants, institutions of higher education could more effectively preserve their independence, integrity, and freedom of inquiry. Institutional grants normally are awarded on the basis of the volume of federally supported research conducted at the institution. Therefore, in the view of its champions, the institutional grant approach offers three key advantages: (1) It rewards quality (since research grants are awarded competitively), (2) It provides flexibility to the institution in offsetting the fragmenting effects of a heavy commitment to project research, (3) It helps build up the capacity of the institution as a whole.

Thus far, the federal government has not seen fit to expand its use of institutional grants. In fact, funding for such support has declined in recent years. Some federal officials feel that in the past, institutional support resulted in the overproduction of scientists in certain fields. They also suspect that universities would be better positioned to cater to their own immediate desires while slighting investigations that might help solve national problems.

Proponents of institutional grants aim to lessen the degree of federal control and direction over academic research activities. In contrast to these views, one university social scientist, Amitai Etzioni, has strongly championed more federal direction of its sponsored research. He would prefer emphasizing contracts rather than grants. He felt that federal officials should specify research goals, ensure that the overall research program is rational and coordinated, and reduce the unauthorized diversion of research effort from stated objectives to those preferred by the performing scientists.[13]

The issue of how much authority the federal government should give to university administrators in controlling federally funded academic research will remain a matter of controversy for years to come. At

---

[13] Etzioni, "Redirecting Research Dollars," p. B-3.

present, federal officials do not seem anxious to grant greatly expanded authority through the mechanism of institutional grants.

Defense-sponsored Research

No problem triggered more emotional outbursts on American campuses than the Department of Defense's sponsorship of academic research. Since World War II the defense establishment has supported such research with enthusiasm. In return, university scientists have made major contributions to national security. However, beginning in the late 1960s, critics increasingly questioned whether universities should perform such research. They charged that: (1) Research for war is immoral, (2) The Defense Department dominates university research, (3) Secret research is incompatible with a university's and science's need for full openness.

Critics argued that since war can lead to loss of life, war is immoral. Scientists can neither ignore nor absolve themselves of the social consequences of their work. Thus, if their efforts aid a nation to wage war, these efforts likewise fall beyond the permissible. Disillusionment with the Vietnam war stimulated a growing chorus that echoed this line. In the late 1960s, objections to defense research became a rallying cry for student riots at Harvard, Stanford, and other schools. Moreover, these critics argued that universities should aid in redirecting the nation's priorities from defense to domestic concerns, and, therefore, should stress "peaceful" research.

Backers of university participation in defense research countered that since universities and scientists alike have a vital interest in promoting the national security of the United States, they should behave like any good citizen. Since universities possess certain unique scientific capabilities, our national security demands that the Defense Department enlist their help. Hopefully, the weapons and equipment that might eventually result from academic research will help to deter rather than fight wars. The Defense Department must remain in contact with the academic scientific community which performs much of the country's basic research.

The critics felt that defense domination of university research has distorted the emphasis that universities accord to the various scientific disciplines and projects. Specifically, only those that give promise of someday producing weapons receive sufficient funding. As a result, they contended, nondefense-related disciplines and projects have suffered.

The Defense Department not only countered with national security arguments, but denied that it dominated. It pointed out that non-defense agencies have become the major federal supporter of research funds. In 1952, the Defense Department contributed $100 million for research, accounting for 73 percent of the federal total. In 1972, it funded $205 million, but this figure represented only 11 percent of the federal total. Defense officials claimed that 11 percent could hardly be called domination. In addition, at the basic research stage of R&D, it is almost impossible to determine when findings ultimately will promote military or civilian purposes.

There is little doubt that an incompatibility exists between the need for openness and secret defense research performed on the campus. Science advances through a dialogue in which scientists have the opportunity to judge the work of their colleagues. If secrecy interferes with this dialogue, it violates a basic tenet of the scientific ethos. Thus, critics at the University of Pennsylvania successfully objected to the secret chemical and biological warfare research carried on by one of the university's research institutions. Other universities also have terminated classified research.

Defense leaders contended that while secret research did prevent full disclosure, freedom of choice still remains a cardinal principle of our democratic society and must be guaranteed in research and education as well as in other sectors of national life. If individual scientists or institutions want to take on secret defense research, they have a right to do so. In some cases, the Department alone supported certain lines of inquiry in which some scientists have an interest, and a few of these must remain secret. Besides, they argued, over time the results that flow even from secret research can become public and can lead to peaceful as well as military uses.

Moreover, the Defense Department contends that it sponsors very little secret research at universities. In 1971, only about 3 percent of total defense-supported academic research was "classified." No doubt the reluctance of universities to permit secret research contributed to this very low figure. For the most part, the end product of defense-financed research is open publication of results. Thus, even if one accepts the argument that secret research has no place on the campus, the defense establishment maintains that very little defense-sponsored research today would be excluded for this reason.

Here again, little change can be expected in the foreseeable future. Most likely, universities, many of whom are presently suffering acute financial difficulties, will not turn away defense-sponsored open research. The matter of secret projects will continue to trouble many, but its share of total defense funding probably will not increase from its present very small size.

Foreign Competition

From the end of World War II to the 1960s the United States had a commanding lead across a wide spectrum of science and technology. Europe and Asia expended most of their energies just recouping from the devastation of World War II. The United States poured much treasure and technology into helping these nations get on their feet. With U.S. help, but mostly by their own efforts, Western Europe and Japan achieved phenomenal economic and technological progress. Between 1965 and 1970 Japan led the world with a 14.2 percent increase in output per man-hour (productivity); West Germany experienced a 5.3 percent gain, while the United States exhibited a disappointing 2 percent gain.

There is a clear correlation between improvements in productivity and the international competitive position of an industry or an entire economy. Part of the success story of Japan and West Germany can be attributed to technological infusions that they received from the United States. However, these countries also set about building up their own technological capabilities. While R&D spending as a share of GNP in the United States was declining in the 1960s, other industrialized countries expanded their efforts. In 1972, West Germany and Japan spent a higher proportion of their GNP on R&D than did the United States. Since neither country spent as much on defense R&D, they probably expended more per capita on civilian research than the United States. They simultaneously made heavy investments in new plant and equipment. Many countries encouraged joint industrial R&D ventures, often subsidizing them; in the United States antitrust laws prohibited such activities.

West Germany allows a depreciation of up to 150 percent of the original cost of R&D capital. Japan permits accelerated depreciation for investment in research and in the plant and equipment to make research economically productive. This emphasis on civilian R&D, on hard work, and on lower labor costs, allowed exports from West Europe and Japan to burgeon to the United States and around the world.

American industrial products increasingly met fierce competition from abroad, driving some American exports from world markets. It even became uneconomical to manufacture many types of electronic components in the United States (although this country had pioneered in their development). Japan, and later Taiwan, Korea, and Hong Kong, became the prime suppliers of these items. Car imports to the United States captured some 20 percent of the American automobile market by 1970. As a result, by 1972, the United States had experienced 6 years of declining trade surpluses, a dollar crisis, and the first trade deficit that this country suffered since 1888. Since 1972, a series of circumstances, including a revalued dollar, enabled the United States to regain a favorable trade balance. The foreign exchange requirements of the grossly inflated price of imported oil could easily reverse this favorable trend.

In order for the United States to maintain an acceptable balance of trade, it had to expand exports in technologically intensive products and services that flowed from its possession of the world's most potent research and development base. It continued to keep technological leadership in areas like computers and aerospace systems, especially in commercial aircraft, and these items continued as major elements of the nation's exports. Moreover, a technological revolution on American farms permitted the increased export of agricultural products at competitive prices. Yet, our national leaders recognized the growing technological challenge from abroad. In 1972, President Nixon pointed out that, "Our international position in fields such as electronics, aircraft, steel, automobiles and shipbuilding is not as strong as it once was. A better performance is essential to both the health of our domestic economy and our leadership position abroad."[14]

In brief, the President asked the country to push those technological

---

[14] Nixon, *The Importance of Our Investment in Technology,* 1972.

frontiers that would make U.S. products more competitive with foreign imports at home and in world markets. The United States had to put more effort into innovative activities that would boost productivity, reducing unit costs of American products, and, thereby, making them more economically attractive. There is little doubt that in the foreseeable future this country will continue to confront fierce foreign competition in technology, especially from Western Europe and Japan. Some economic and technological accommodations among these trading nations seem inevitable. In addition, statesmen are looking for areas of cooperation that could dampen tendencies toward cutthroat technological rivalry. This task imposes enormous challenges.

Détente and Technology Transfer

By 1972 President Nixon had gone far in moving from an era of confrontation to one of negotiation. His trips to the Soviet Union and to the People's Republic of China heralded the new age. During his trip to Moscow the President signed a series of agreements, including some dealing with the Strategic Arms Limitation Talks I (SALT I) and with cooperation in science and space, designed to improve relations between these two superpowers. The Soviets have opted for détente for many reasons, but surely one is their desire to gain access to advanced technology in the West, and especially in the United States. The general problem of technology transfer to foreign nations presents the United States with some difficult problems (and attractive opportunities). Soviet requests for advanced technology raise some even more perplexing questions since they touch on matters of national security as well as economics. Will supplying high technology contribute directly to the Soviets' warmaking capacity or ease their economic burden, thereby permitting them to shift their limited resources from domestic to military pursuits? Should the United States insist that the Soviets redirect their priorities from military to peaceful objectives before granting a major increase in technology transfer? A continuing debate has taken place in Washington, for instance, concerning the wisdom of easing restrictions on the export of computers. This debate has centered on the question of the degree to which such technology transfer would either directly or indirectly improve the Soviets' military posture. At least in one case, President Nixon decided that the potential for increased military capability was not strong enough to deny selling the Russians certain research computers.

Some critics also have expressed concern over the economic consequences of such transfers. They fear that the Soviets might only buy prototypes or production knowledge and then go into manufacturing on their own. If they take this route, the United States would trade away a great deal of advanced technology on a one-shot basis, gaining little in return. On the other hand, the potential for expanded East-West trade through technology transfer looks attractive to many American industrialists. Above all, the Soviets must have some incentive for pursuing détente, and the hope of improving their technological capacity seems an incentive with which the United States can live and perhaps even benefit.

No doubt solutions to some of these problems will be worked out over time. Unless the political situation deteriorates, in the foreseeable future it appears likely that advocates of accelerated technology transfer to the Soviet Union will enjoy some success. However, government officials will judge the merits of such transfers on a case by case basis.

Summary

While science has helped mankind, it also has given him some potent problems. The vastly increased role of the federal government in the nation's science and technology enterprise likewise has both facilitated gain and caused difficulties. In the area of federal decisionmaking, the role of science advisers throughout the governmental structure commands our attention. We require a better understanding of the dynamics of selecting advisers (especially part-time consultants), their missions and responsibilities, their need to communicate clearly and intelligently with policymakers and with the public. Allocating resources ever remains an exacting task, and, given the overlapping activities of mission-oriented agencies, coordinating their R&D programs presents no less a challenge.

In regard to general substantive problems, man must constantly confront the consequences of rapid change. While recognizing that economic and technological growth, depersonalization and antitechnology attitudes cause serious problems, man would be foolish to turn against science and technology. Instead, he should strive to exert social control over them and their consequences as well as to design tools like technology assessment to help him in this task.

The specific substantive issues discussed here illustrate the great efforts needed to ensure a healthy national science and technology program. Those responsible for allocating resources no doubt will continue to find it difficult to please simultaneously the advocates of basic versus applied research. The entire problem of academic research has become more serious in recent years. Cutbacks in federal support for this type of R&D require that universities manage their affairs with increased effectiveness. The problem of defense-sponsored research on campuses will continue to trouble some, but, most likely, university scientists will continue to seek and receive defense funds. University administrators will continue to ask the federal government to shift emphasis from project to institutional support; the degree of their success remains an open question.

Foreign competition poses a serious challenge to our scientists and technologists to increase this country's productivity, create a demand for new products, and improve the foreign trade position of the United States. Lastly, the issue of technology transfer to foreign nations in general has commanded increasing attention while the export of advanced technological knowledge to the Soviet Union under conditions of détente has become a point for intensive debate in this country.

Suggestions for Further Reading

Barber, Bernard. *Science and the Social Order.* New York: Collier Books, 1962.

————, and Walter Hersch (eds.). *The Sociology of Science.* New York: Free Press of Glencoe, 1962.

Bright, James R. *Automation and Management.* Boston: Division of Research, Graduate School of Business Administration, Harvard University, 1958.

Brooks, Harvey. *The Government of Science.* Cambridge, Mass.: The MIT Press, 1968.

Brown, Fred R., and Stephen B. Chitwood. Highlights from the Literature on Organization for Federal Programs in Science and Technology. Washington, 1968. Microfiche.

Bush, Vannevar. *Science: The Endless Frontier.* Washington, D.C.: Public Affairs Press, 1946.

Cohen, Morris R., and Ernest Nagel. *An Introduction to Logic and the Scientific Method.* New York: Harcourt, Brace, 1934.

Commoner, Barry. *Closing Circle: Nature, Man, and Technology.* New York: Knopf, 1971.

Conant, James B. *On Understanding Science: An Historical Approach.* New Haven: Yale University Press, 1947.

Conference Board, The. *Information Technology: Some Critical Implications for Decisionmakers.* New York: The Conference Board, 1972.

Dupre, Joseph S., and Sanford A. Lakoff. *Science and the Nation: Policy and Politics.* Englewood Cliffs, N.J.: Prentice-Hall, 1962.

Duprée, A. Hunter. *Science in the Federal Government.* Cambridge, Mass.: Harvard University Press, 1951.

Ellul, Jacques. *The Technological Society.* John Wilkinson (trans.). New York: Knopf, 1964.

Gilpin, Robert. *American Scientists and Nuclear Weapons Policy.* Princeton: Princeton University Press, 1962.

————, and Christopher Wright (eds.). *Scientists and National Policy-Making.* New York: Columbia University Press, 1964.

Glaser, Barney G. *Organizational Scientists: Their Professional Careers.* New York: Bobbs-Merrill, 1965.

Hailsham, Quintin McG. H., 2d Viscount. *Science and Politics.* Chicago: Encyclopedia Britannica Press, 1964.

Haskins, Caryl P. *The Scientific Revolution and World Politics.* Elihu Root Lectures, 1961-1962. New York: Harper and Row, 1964.

Hill, Karl B. (ed.). *The Management of Scientists.* Boston: Beacon Press, 1964.

Hitch, Charles J., and Roland N. McKean. *The Economics of Defense in the Nuclear Age.* Cambridge, Mass.: Harvard University Press, 1960.

Jewkes, John, David Sawers, and Richard Stillerman. *The Sources of Invention.* London: Macmillan and Co., 1960.

Jordan, Robert S. (ed.). *Multinational Cooperation: Economic, Social, and Scientific Development.* New York: Oxford University Press, 1972.

Kaplan, Norman (ed.). *Science and Society.* Chicago: Rand McNally, 1965.

Kast, E. Fremont, and James E. Rosenzweig (eds.). *Science, Technology, and Management.* New York: McGraw-Hill, 1963.

Knorr, Klaus, and Oskar Morgenstern. *Science and Defense: Some Critical Thoughts on Military Research and Development.* Princeton: Center of International Studies Woodrow Wilson School of Public and International Affairs, Princeton University, 1965.

Kuhn, Thomas S. *The Structure of Scientific Revolution.* Chicago: University of Chicago Press, 1962.

Lapp, Ralph E. *The New Priesthood.* New York: Harper and Row, 1965.

Lindsay, Robert B. *The Role of Science in Civilization.* New York: Harper and Row, 1963.

Lindveit, Earl W. *Scientists in Government.* Washington, D.C.: Public Affairs Press, 1960.

McCamy, James L. *Science and Public Administration.* Montgomery: University of Alabama Press, 1960.

Margulis, Newton, and Anthony P. Raia (comps.). *Organizational Development: Values, Process, and Technology.* New York: McGraw-Hill, 1971.

Mees, C. E. Kenneth, and John A. Leermakers. *The Organization of Industrial Research.* New York: McGraw-Hill, 1950.

Michael, Donald. *Cybernation.* Santa Barbara, Calif.: Center for the Study of Democratic Institutions, 1962.

National Academy of Sciences. *Basic Research and National Goals: A Report.* Report for U.S. Congress, House, Committee on Science and Astronautics, 89th Cong., 1st sess. (Washington, D.C.: U.S. Government Printing Office, 1965.

National Industrial Pollution Control Council. Series of staff and sub-council reports. Washington, D.C.: U.S. Government Printing Office, 1970-1972.

National Research Council. *Criteria for Federal Support of Science.* Panel discussion, March 11, 1968. Washington, D.C.: National Academy of Sciences, 1969.

Nelson, Richard B. *Science, the Economy, and Public Policy.* Santa Monica, Calif.: Rand Corporation, 1964.

Newman, James R. *What Is Science?* New York: Simon and Schuster, 1955.

Oliver, John W. *A History of American Technology.* New York: Ronald Press, 1956.

Ozbekhan, Hasan. *Technology and Man's Future.* Santa Monica, Calif.: System Development Corp., 1966.

Price, Derek J. de S. *Little Science, Big Science.* New York: Columbia University Press, 1963.

Price, Don K. *Government and Science.* New York: New York University Press, 1954.

Raudsepp, Eugene. *Managing Creative Scientists and Engineers.* New York: Macmillan, 1963.

Reagan, Michael R. *Science and the Federal Patron.* New York: Oxford University Press, 1969.

Schilling, Warner R. "Scientists, Foreign Policy, and Politics." *American Political Science Review,* 56, June 1962.

Seaborg, Glenn T., and William R. Corliss. *Man and Atom.* New York: E. P. Dutton, 1971.

Singer, Charles, et al. (eds.). *A History of Technology.* Vols. III, IV. Oxford: Oxford University Press, 1957, 1958.

Snow, Sir Charles P. *The Two Cultures and the Scientific Revolution.* Cambridge, Mass.: Cambridge University Press, 1962.

Stover, Carl F. *The Government of Science.* Santa Barbara, Calif.: Center for the Study of Democratic Institutions, 1962.

Syracuse University Research Corporation. *United States and the World in the 1985 Era, with Appendix* (committee reports). Syracuse, N.Y., 1964.

U.S. Congress, House, Committee on Government Operations. *The Federal Research and Development Programs: The Decisionmaking Process.* Hearings. 89th Cong., 2d sess. Washington, D.C.: U.S. Government Printing Office, 1966.

———, Committee on Science and Astronautics. *Government and Science.* Committee Print. 88th Cong., 1st sess. Washington, D.C.: U.S. Government Printing Office, 1963.

———, Select Committee on Government Research. *Federal Research and Development Programs.* Hearings, Part 2. 88th Cong. Washington, D.C.: U.S. Government Printing Office, 1964.

———, Subcommittee of the Committee on Government Operations. *Organization and Administration of Military Research and Development Programs.* Hearings. 83d Cong., 2d sess. Washington, D.C.: U.S. Government Printing Office, 1954.

U.S. Department of Commerce. *Science and Technology for Mankind's Progress.* Washington, D.C.: U.S. Government Printing Office, 1967.

U.S. National Resources Committee. *Technological Trends and National Policy.* Washington, D.C.: U.S. Government Printing Office, 1937.

U.S. National Science Foundation. *Organization of the Federal Government for Scientific Activities.* Washington, D.C.: U.S. Government Printing Office, 1956, 1961.

U.S. President's Science Advisory Committee. *Science, Government and Information.* Washington, D.C.: U.S. Government Printing Office, 1963.

Walker, Charles R. *Modern Technology and Civilization.* New York: McGraw-Hill, 1962.

Weiner, Norbert. *Cybernetics.* Cambridge, Mass.: M.I.T. University Press, 1948.

Whitehead, Alfred N. *Science and the Modern World.* New York: Macmillan, 1964.

Wiesner, Jerome B. *Where Science and Politics Meet.* New York: McGraw-Hill, 1965.

Wolfe, Dael. "Government Organization of Science." *Science,* 131, May 13, 1960.

Index

137